Culture and the Environment in the Himalaya

This book is concerned with human–environment relations in the Himalaya. It explores how different populations and communities in the region understand or conceive of the concept of environment, how their concepts vary across lines of gender, class, age, and status, and what this implies for policy-makers in the fields of environmental conservation and development. The chapters in this book analyze the symbolic schema that shape human–environment relations, whether that of scientists studying the Himalayan environment, public officials crafting policy about it, or people making a living from their engagement with it, and the way that natural phenomena themselves shape the human perception of the world. A trans-global environmentalism that sees human actions as reshaping the world in negative ways must come to terms with other cultural perspectives that do not share that view. As a new approach to the study on the environment in South Asia, *Culture and Environment in the Himalaya* introduces the new thinking in environmental anthropology and geography into the study of the Himalaya and uses Himalayan ethnography to interrogate and critique contemporary theorizing about the environment.

Arjun Guneratne is Professor of Anthropology at Macalester College, USA, and author of *Many Tongues, One People: The Making of Tharu Identity in Nepal*. He is the editor of *Himalaya*, the journal of the Association for Nepal and Himalayan Studies.

Routledge Contemporary South Asia Series

1 Pakistan
Social and cultural transformations in a Muslim nation
Mohammad A. Qadeer

2 Labor, Democratization and Development in India and Pakistan
Christopher Candland

3 China–India Relations
Contemporary dynamics
Amardeep Athwal

4 Madrasas in South Asia
Teaching terror?
Jamal Malik

5 Labor, Globalization and the State
Workers, women and migrants confront neoliberalism
Edited by Debdas Banerjee and Michael Goldfield

6 Indian Literature and Popular Cinema
Recasting classics
Edited by Heidi R.M. Pauwels

7 Islamist Militancy in Bangladesh
A complex web
Ali Riaz

8 Regionalism in South Asia
Negotiating cooperation, institutional structures
Kishore C. Dash

9 Federalism, Nationalism and Development
India and the Punjab economy
Pritam Singh

10 Human Development and Social Power
Perspectives from South Asia
Ananya Mukherjee Reed

11 The South Asian Diaspora
Transnational networks and changing identities
Edited by Rajesh Rai and Peter Reeves

12 Pakistan–Japan Relations
Continuity and change in economic relations and security interests
Ahmad Rashid Malik

13 Himalayan Frontiers of India
Historical, geo-political and strategic perspectives
K. Warikoo

14 India's Open-Economy Policy
Globalism, rivalry, continuity
Jalal Alamgir

15 The Separatist Conflict in Sri Lanka
Terrorism, ethnicity, political economy
Asoka Bandarage

16 India's Energy Security
Edited by Ligia Noronha and Anant Sudarshan

17 Globalization and the Middle Classes in India
The social and cultural impact of neoliberal reforms
Ruchira Ganguly-Scrase and Timothy J. Scrase

18 Water Policy Processes in India
Discourses of power and resistance
Vandana Asthana

19 Minority Governments in India
The puzzle of elusive majorities
Csaba Nikolenyi

20 The Maoist Insurgency in Nepal
Revolution in the twenty-first century
Edited by Mahendra Lawoti and Anup K. Pahari

21 Global Capital and Peripheral Labour
The history and political economy of plantation workers in India
K. Ravi Raman

22 Maoism in India
Reincarnation of ultra-left wing extremism in the 21st century
Bidyut Chakrabarty and Rajat Kujur

23 Economic and Human Development in Contemporary India
Cronyism and fragility
Debdas Banerjee

24 Culture and the Environment in the Himalaya
Arjun Guneratne

Culture and the Environment in the Himalaya

Edited by
Arjun Guneratne

Routledge
Taylor & Francis Group
LONDON AND NEW YORK

First published 2010
by Routledge
2 Park Square, Milton Park, Abingdon, Oxfordshire OX14 4RN

Simultaneously published in the USA and Canada
by Routledge
711 Third Avenue, New York, NY 10017

First issued in paperback 2012

Routledge is an imprint of the Taylor & Francis Group, an informa business

Typeset in Times by RefineCatch Limited, Bungay, Suffolk

British Library Cataloguing in Publication Data
A catalogue record for this book is available
from the British Library

Library of Congress Cataloging-in-Publication Data
Guneratne, Arjun, 1961–
Culture and the environment in the Himalaya/Arjun Guneratne.
p. cm.—(Routledge contemporary South Asia series)
Includes bibliographical references and index.
1. Human ecology—Himalaya Mountains. 2. Forest ecology—Himalaya Mountains. 3. Forest conservation—Himalaya Mountains. 4. Environmental change—Himalaya Mountains. 5. Himalaya Mountains—Environmental conditions. I. Title.
GF696.H55G86 2010
304.2095496—dc22
2009022559

ISBN13: 978-0-415-77883-1 (hbk)
ISBN13: 978-0-415-53314-0 (pbk)
ISBN13: 978-0-203-86436-4 (ebk)

This book is for my children, Ananda and Sara

Contents

List of illustrations

Figures

Tables

Preface and acknowledgments

This book originated out of my interest in addressing the relative lack of attention paid to the issue of cognized environments in the Himalaya, a region with an otherwise rich literature on the environment, by posing the general question of how different cultures in the Himalaya (whether those of bureaucrats, villagers, development workers, ethnic groups or others) conceptualized the environment. Most of the chapters collected here were originally presented as papers at a conference at Macalester College in the summer of 2004, and subsequently revised; the chapters by Fortier and Cameron were included later, as the volume was being prepared. Emma Mawdsley's chapter was published in *Worldviews*, and is reprinted here with minor revisions. The chapters—contributed by anthropologists and cultural geographers—are mostly concerned with Nepal, but the chapters by Mawdsley and Aggarwal deal with the Indian Himalaya, and Subba discusses the ethnology of the Limbu, who straddle the border between Nepal and Sikkim. They also vary in terms of the cultural scale on which the analysis is pitched—ranging from the close examination of the ideas of an individual (Rademacher) to an ethnic group (Fortier, Subba, Russell) to a scientific community (Metz, Cameron)—but all of them, in different ways, are concerned to examine what exactly environment means in the various contexts in which the authors have worked. They do so through the examination of a number of themes that are discussed at length later in the Introduction.

The conference was made possible by the generous assistance of the President of Macalester College, Brian Rosenberg, and the then provost, Dan Hornbach, who provided funds that allowed a number of the participants to travel to St. Paul. Tom Robertson and David Gellner were helpful in suggesting people who might contribute to this project, and Mary Lou Byrne provided considerable help in organizing the conference. Birgit Mühlenhaus produced the map at very short notice. My colleagues in the Department of Anthropology—Jack Weatherford, Dianna Shandy, and Sonia Patten—have been a great source of support, and my wife, Kate Bjork, gave me critical feedback on the Introduction to the volume. I am deeply grateful to all of them. I also want to thank my editor at Routledge, Dorothea Schaefter, for her confidence in this project.

Contributors

Safia Aggarwal is a geographer, and works with ARD, Inc. in Burlington, Vermont, as a land tenure and natural resource specialist, providing technical support in the design, implementation, monitoring and evaluation of international programs in natural resource governance, biodiversity conservation, decentralization and devolution, and land tenure and property rights reforms. She obtained her doctoral degree in geography from the East-West Center and the University of Hawai'i where she conducted research on institutions of commons management in the mid-hills of Kumaun, India.

Mary Cameron is Associate Professor of Anthropology in the Department of Anthropology at Florida Atlantic University in Boca Raton, Florida. Her research focuses on gender and caste relations in Nepal's farming communities, the evolution of dalit identity in post-revolution Nepal state formation, and medicine. Her current research on medicine examines the impact of western science and biomedical modernization on Ayurvedic medicine and medical pluralism in Nepal, and women Ayurvedic doctors and healers.

Ben Campbell is a lecturer in anthropology at Durham University, UK. He has conducted extensive fieldwork with Tamang-speaking communities in Nepal on social relations of farming and environmental practice; and as a member of a Europe-wide project has written on the public response to genetically modified foods.

Jana Fortier teaches and does cultural resource management consulting in San Diego. She previously was a tenured Associate Professor of Anthropology at Southwest Minnesota State University. Her field research focuses on the dynamics of cultural survival among Himalayan foragers known as Raute and Ban Raji.

Arjun Guneratne is Professor of Anthropology at Macalester College in St Paul, Minnesota. He has worked in Nepal on the relationship of state formation to ethnic identity among the Tharu of the Tarai. His current research examines the relationship of biodiversity conservation in Sri Lanka to neo-liberal economic reforms and globalization. He is the editor of *Himalaya*, the journal of the Association for Nepal and Himalayan Studies.

Emma Mawdsley is a Lecturer in Geography at Cambridge University, UK, and a Fellow of Newnham College. She has worked on issues of environment, development and regionalism in Uttarakhand, India; her other environment-focused research includes analyses of the Hindu Right, and the environmental beliefs and values of India's middle classes. She is also interested in global development politics more broadly, and is presently leading a British Academy-funded project on contemporary India–Africa relations.

John J. Metz spent three years as a Peace Corps Volunteer in Iran after earning a philosophy undergraduate degree. After Iran, he worked in a variety of social service jobs for five years before entering a Geography PhD program at the University of Wisconsin. This led him to Nepal to study how the farming system of an upper elevation village in west Nepal utilized and affected the local forest vegetation. In addition to interests in human use of Himalayan environments, he has also studied the roles, successes, and failures of citizen advisory boards in the "clean-up" of the US nuclear weapons factories. He teaches Geography at Northern Kentucky University in suburban Cincinnati.

Andrea Nightingale is a Lecturer in Environmental Geography at the University of Edinburgh in Scotland. She has worked in Nepal on issues of environmental change and social justice in Community Forestry. Her current research focuses on the emergence of democracy and civil society practices in relation to natural resource management and post-conflict politics.

Anne M. Rademacher is an Assistant Professor in the Programs in Environmental Studies and Metropolitan Studies at New York University. An environmental anthropologist with research interests in urban sustainability and the cultural politics of nature, her work in Kathmandu explores contests over urban ecology in the context of democratization, political emergency, and state transformation.

Andrew Russell is Senior Lecturer in the Department of Anthropology at Durham University, UK. His research in Nepal concerns environment, identity and migration among the Yakkha. His other research interests are in public health, ecotourism and cross-cultural education. He is currently Director of Phoenix, Durham's community outreach and engagement programme.

T.B. Subba is Professor of Anthropology at North-Eastern Hill University, Shillong, India. His research interests include ethnicity, development, culture, and diasporic studies. He has done intensive fieldwork in Darjeeling, Sikkim and East Nepal. He edits a multi-disciplinary and internationally refereed journal called *The NEHU Journal* (www.nehu.ac.in). He has held various administrative positions at North-Eastern Hill University.

Note to the reader

Hindi and Nepali words have been transliterated following the convention in R.L. Turner's *A Comparative and Etymological Dictionary of the Nepali Language*. Thus, 'c' is pronounced 'ch' (e.g., *panchāyat*, which is spelled, following Turner, as *pancāyat*), while aspirations are indicated with an 'h' (e.g., *śobhā*). The transliteration of other words in Asian languages (e.g., Raute, Yakkha, Limbu, Tamang, and Tibetan) is that of the respective authors of the chapters in which they appear. Words in South Asian languages have been italicized, but I have rendered plurals with an 's' rather than use the original plural form. Place names and the names of ethnic groups are rendered in conventional English spelling and are not in italics.

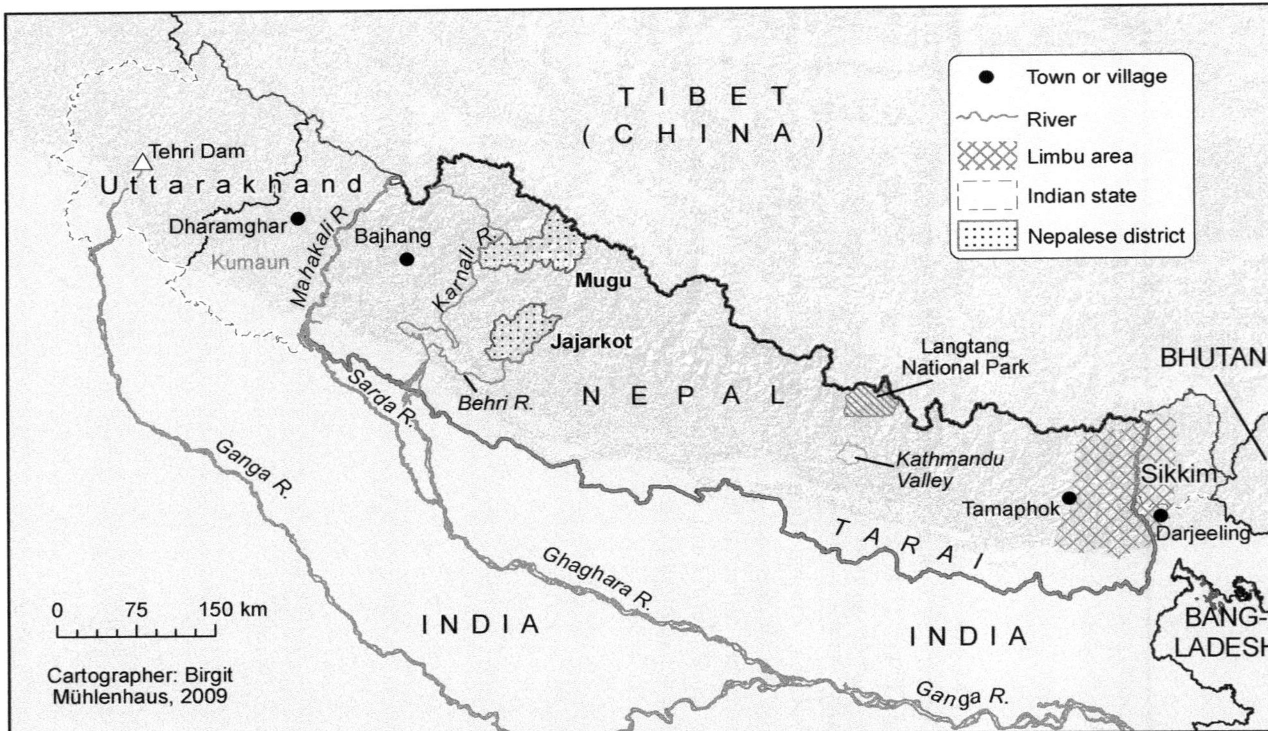

Map of the Central Himalaya, showing areas mentioned in the text.

Introduction

Arjun Guneratne

Although recent scholarship has yielded a rich literature on the impact that human activity has had on the Himalayan environment and the uses which human beings have made of it, there has been relatively little attention paid to how human beings who live and work in that part of the world understand or conceptualize it, and the implications of their cognized models of the environment for policy-making in development work.[1] The authors of the collection of essays presented in this book seek to expand our knowledge of the relation between human beings and their environment by focusing on the cognized environment, and in so doing to direct attention to areas that are often ignored in the study of the Himalaya. Among Himalayan states, Nepal in particular has been consumed by the ideology of development, and scientific views (including out-dated ones) that pay little or no attention to cultural concerns in their analysis of human–environment relations have dominated policy-making by bureaucrats and development workers. For their part, anthropologists and cultural geographers working in the Himalaya have either focused their attention on issues such as ethnicity, kinship, the caste structure, shamanism, religion and the like (cf. Fisher 1985) when they were concerned about "culture," or on the way that Himalayan societies made use of natural resources and the impact of their activities on nature when they took human–environment relations as the object of their study.

What environment means to people who live and work in the Himalaya, and the impacts of those understandings are the subject of this book. Environment is a concept constituted in relation to the person whose environment it is, mediated by the symbolic system through which an individual apprehends and gives meaning to the world. "Environment" contains not just bio-physical but also socio-cultural components, including other human beings and *their* cognized environments. We discuss in this book how Himalayan communities conceptualize their environment, and how their concepts vary across lines of gender, class, age, status, ethnicity, level of and kind of education and other forms of social experience.

The dominant perspective on the Himalayan environment, which crystallized during the decade of the 1970s and has endured in policy-making circles ever since has been referred to as the Theory of Himalayan Environmental

Degradation (known to those who care about it as THED). This view of human–environment relations may be traced to the dismal perspective on the human condition first articulated in the eighteenth century by the English demographer Thomas Malthus. Malthus did not subscribe to the optimism of his contemporaries regarding the perfectibility of humankind. He believed rather that the increase in the human population—by which he meant the increase in the population of the lower classes, to whom he attributed much of the blame for human fecundity—would outstrip the growth in the food supply, and lead to widespread famine and the degradation of the world's resources.

Beginning in the 1970s, scientists and policy-makers connected flooding on the Gangetic plain and the heavy load of silt borne by rivers originating in the Himalaya to pressures exerted by a rapidly expanding population on the environment in the middle hills of Nepal. They concluded that the pressure of a growing population on mountain resources had generated a crisis of deforestation, soil erosion and general environmental degradation. While this argument held sway throughout the 1970s, it has been challenged in the decades since by scholars who argue that it was greatly oversimplified, ahistorical and based on inadequate research (Ives, 1987; Metz, Chapter 1 in this volume).

While THED has provided the dominant paradigm for understanding human–environment relations in the Himalaya, a second significant approach has focused on the ways in which human populations adapt to and utilize the environment. It has been less concerned with the question of culture or cognition, except as understood in a very limited and utilitarian way. Thus, Müller-Böker and Kollmair, in contrasting "traditional" societies to "modern" ones, write: "Concepts of nature in traditional societies differ from those of 'modern' societies mainly because the structures of production are fundamentally different. Their evaluation of the natural environment is primarily utilization-based" (2000: 325). While the premise of this assertion may be questioned (surely modern industrial society is no less utilitarian in the way it views the world), such studies, not surprisingly, tend to focus on ethnoecological knowledge and resource use (e.g., Daniggelis 1997; Müller-Böker 1999). They do not directly address the questions posed in this volume; they do not ask how their subjects conceive of the concept of environment or of nature but rather ask what knowledge their subjects possess of an environment already defined by the researcher in bio-physical terms. Müller-Böker's work on the Tharu (1999) is an excellent example of this genre; focusing on ethnoecological knowledge yields a detailed compendium of information on plants and soils and local understandings of their attributes, but it tells us little of the cognized environment in which Tharus live their lives.

A third approach that has developed in recent years focuses attention on how human beings give meaning to landscapes and is closer to the approach advocated in this book. In anthropology generally, which has been a latecomer to treating landscape as a concept worthy of analytical attention (geographers and art historians were first on the scene), the emphasis has been on exploring the validity of such a concept in other cultures and in inquiring into how landscapes are produced through practice and social memory (Bender 1993; Hirsch and O'Hanlon 1995). In the ethnology of the Himalaya, this enterprise has been pursued with reference mainly to sacred landscapes and with an emphasis on pilgrimage (Huber 1999; Gutschow *et al.* 2003; Sax 2003; Arora 2006). This approach—of asking whether 'landscape' as a concept has cross-cultural meaning and if so, where that meaning lies—has the most affinity with the project we pursue in this book.

The dimensions of culture

Human beings live suspended in webs of significance that they themselves have spun, to borrow a metaphor from Clifford Geertz (1973: 5). They do not spin those webs in splendid isolation from their material activity, however; the reproduction of the physical existence of individuals, as Marx put it, is the product of particular forms of activity to which have been attributed specific kinds of meanings. To quote Marx, "The phantoms formed in the human brain are also, necessarily, sublimates of their material life process, which is empirically verifiable and bound to material premises" (Marx and Engels 1970: 47). These symbolic schema, whether they inform the actions of scientists studying the Himalayan environment, of planners seeking to develop it or villagers striving to make a living from it, have consequences for the way human action shapes bio-physical processes and the way that natural phenomena themselves shape human perception of the world. A trans-global environmentalism that sees human actions as reshaping the world in ways that it considers detrimental must come to terms with other cultural perspectives that do not share that view (cf. Milton 1996; Alley 2002).

Given that cultures are not unitary, homogeneous phenomena, universally shared by all members of a society, it might reasonably be asked whether we can speak of cognized environments in any meaningful sense. The alternative, however, is a radical particularism at the level of the individual, which is equally untenable. As John and Jean Comaroff have pointed out, despite our sensitivity to the problem of the reification of culture, no one doubts that such a thing as "Balinese culture" exists (Comaroff and Comaroff 2006: 383). Similarly, we can think about something like Tamang culture or Limbu culture while acknowledging that culture is, indeed, as the Comaroffs put it "a set of polyvalent practices, texts, and images that may, at any time be contested" by others in the social field in which those practices, texts, and images are manifest (ibid.: 384).

I draw on two concepts to reconcile the notion of a cognized environment with the fact that culture is not a unitary or homogeneous set of meanings: that of communication and of cultural scales. The notion of cultural scale refers to the extent of the social field in which actors engage each other in sustained ways in the process of reproducing themselves and their societies.

That engagement—which I will refer to as communication—produces over time a shared language and a common set of meanings, which are continuously interrogated and transformed as a function of the particular experiences of individual actors. A consensus or agreement on some things at least is necessary for human beings to act in the world; that is the basis of language and other symbolic forms, whose meanings are not fixed for all time but open to contestation, interpretation and transformation as a function of human experience. A society that cannot achieve a fundamental consensus about how the world is to be symbolically coded would be unworkable. We call such a consensus culture, and we understand the consensus to vary depending on the scale of the society whose consensus it is.

Sustained communication over a period of time by members of a community leads to a more or less shared understanding and set of assumptions of the way the world is; if it did not, the community or the group would soon cease to exist as such. The smaller the social field, the more intense is the communication among the actors, and the greater the degree of consensus is likely to be. The broader the social field, the weaker the consensus (*doxa* in Bourdieu's (1977) terms); in such a context, one is likely to find emergent orthodoxies rather than a more or less stable doxic order. Another way of putting this is to say that while culture is a dynamic consensus or agreement on what things mean or should mean, arrived at over time by actors engaged with each other in the on-going processes of social life, the contours of that consensus will vary across different sub-groups in the same community and even across individuals as a result of their social experience. As Keesing (1987) reminds us, cultural meanings are socially situated meanings, and knowledge of the environment is differentially distributed across a population. Culture as a system of shared symbols is grounded in the persistent, enduring and structured relations that link human beings to each other and to their need to make the world work for them.

We might expect to see a high degree of consensus over meanings among a small band of foragers, such as the Raute described by Fortier in Chapter 5 in this book, while members of a large and complex society might share a more limited set of assumptions about the world. We would also expect to find cultural differences between men and women among the Raute, deriving from their different social experiences, but the differences would not be as marked as among men and women in more complex and stratified societies. While the Raute are by no means isolated, and must deal with forces external to their society, they do not appear inclined to exploit alternative lifeways—such as wage labor—that their situation as part of a modern state affords them. This is unlike the Tamang in Campbell's account or the Limbu in Subba's; young people in these societies are seeking out opportunities outside the confines of the village. The broader spheres of communication this creates thus expose the Tamang and Limbu cultural orders (their consensus about social and cultural reproduction) to a greater repertoire of dissent and reformulation and alternative ways of knowing than characterizes the Raute. Thus, the more disparate and heterogeneous the society, the weaker the degree of communication and the weaker the consensus that is arrived at.

Culture is a symbolic scheme, an ordering of concepts and categories through which human beings understand and interpret the world around them. It is cultural because it is shared, and it is shared through communication. Groups of people who live and work in close and intensive interaction are likely to communicate more and thus share more with each other than with more distant outsiders, in terms both of social distance as well as geography. Thus, we can talk about the culture of a particular village or the culture of the women in that village or the culture of the Tamang in the sense of this shared framework of communication (the very fact that we must operate with labels like Tamang, Limbu, and Raute, implies an emic consensus about some set of fundamental meanings peculiar to the group so labeled—otherwise such labels would be entirely meaningless). While Tamang and Brahmans in a particular village will share a culture because they are part of a community bound together by networks of communication, Tamang among themselves will have even more developed and denser networks of communication that will carve out a distinctively Tamang communicative (and thus cultural) space. The same point could be made with reference to Tamang women or children in the local school.

Communicative practice produces a road map to social life. People may or may not choose to follow the roads marked on the map, preferring trails that cut different pathways through the landscape, but the road map nevertheless exists. Fortier's apposite description of differing ways of naming (and conceptualizing) the same toponymic features by Raute foragers and local farmers is an example: the same topography conveys different meanings to different groups based on the particular qualities of their material engagement with the biophysical world, a point made years ago by Mary Douglas in a now classic paper (Douglas 1975). There is limited communication between these groups, but intense communication within them, oriented to their reproduction as two separate communities. Any given individual will occupy positions in a number of social networks and thus share in a number of different cultural schemes. We could talk about the culture of scientists, or development planners who share symbolic schemes through communication and a particular kind of socialization (modern education), and we might discuss how this field of communication intersects with or is informed by other fields of communication of which these people are part. Stacy Pigg, for instance, describes the twin worlds that development workers in Nepal must straddle in terms of two conflicting knowledge bases, the village-based knowledge they grew up with and the knowledge of *bikās* imparted to them by the development projects they work for, and the cognitive (and strategic) need to compartmentalize these two forms of knowledge in their daily lives (Pigg 1993: 52–3). We need not assume that culture is a coherent, integrated whole; rather, cultures are relational and positional, and depending on the scale, exhibit or are marked by varying degrees of discursive unity.

The idea of communication therefore helps us to conceptualize the way that knowledge and perspectives are distributed in a given community characterized by sustained interaction and communication among its members—what Dan Sperber has called an epidemiology of representation (Sperber 2006). Sperber uses the metaphor of epidemiology to argue that culture, which consists of mental representations of things, is distributed across society in much the same way that disease is. Representations vary; some are fleeting and confined to the individual, others are more enduring and shared to varying degrees by many individuals. Unlike disease, representations tend to be transformed every time they are transmitted; the less liable to transformation, the more cultural those representations are. "To explain culture" says Sperber, "is to answer the following questions: why are some representations more successful in a human population, more contagious, more 'catching' than others?" (ibid.: 311).

This is the import of Subba's Chapter 6 in this volume, in which he discusses the way that Limbus conceive of the environment. Presumably, not all Limbus share in every aspect of the worldview that Subba describes; but those Limbus who are part of a communicative context embedded in the countryside will conceive of this worldview as a general truth, whatever their particular take on it. Limbus who have moved to urban areas are less likely to be invested in this cognitive framework or even to have knowledge of it. As Subba argues, it is less important to their lives; consciousness is the product of the way in which social reproduction is carried on, and urban Limbus engage a different material environment (and must therefore make sense of it in different ways) than do their fellows in the village. At the same time, they will continue to have knowledge of and be invested in other domains of Limbu life—such as life cycle rituals, for example—that they will share with Limbus elsewhere, and which will define them as Limbus and make the label meaningful.

Among social groups like the Limbu, we can expect to find sub-groups whose communicative links are likely to be even more pronounced than they are with their larger social contexts; thus Limbu women are likely to have a different perspective on the environment as a function of their particular gender or caste-based engagement with it. This is a point that is underscored by almost all the chapters in this volume. People's status (gender, social class, caste, ethnicity, age and any number of other axes of analysis) shapes the way people understand and make sense of their environment, even as those understandings remain grounded in a shared symbolic structure that is the distillation of generations of communicative action (see especially the chapters by Subba, Nightingale, Aggarwal, Russell and Campbell for a discussion of these issues).

The contemporary globalization of environmental discourse means that pre-existing systems of knowledge and ways of understanding the environment in the Himalaya will be affected in some degree by these external forces, especially when those forces come coupled to the powerful ideological force of *bikās* (the ideology of development that is ubiquitous in Nepal—see Pigg, 1993). Different social strata in local populations are variously open to concepts and ideas circulating trans-nationally as a function of their status in society; the greater their possession of economic and cultural capital, the closer and more receptive they are likely to be to this discourse. Given the power of developmentalist ideology in a country like Nepal (Pigg 1993; Shrestha 1995; Des Chene 1996), such discourses (identified with modern ways of knowing, which is a positive value) serve as markers of social status. THED is one such discourse; its enduring value comes from its utility in providing the rationale for developmentalist interventions of various sorts (Guthman 1997; Metz, Chapter 1 in this volume), and like the anti-politics machine described by Ferguson (1994), it absolves the development planner from having to deal with the structural inequalities of life in Nepal and elsewhere in the Himalaya.

Tim Ingold, whose ideas on the matter are distilled in his 1992 essay, challenges the idea that human engagement with the environment (which is to say, the human life process) is dependent on culture. His ideas are discussed by Campbell in his contribution to this volume. Ingold is probably the most trenchant critic in environmental anthropology of this anthropological truism. He argues that it is possible to discover meaning in the world directly through our actions, unmediated by a conceptual scheme. In place of the idea that human beings apprehend their environment through the conceptual schemes of their culture, Ingold wishes to substitute what he refers to as the mutualism of person and environment. According to him, the organism "fits the world to itself, by ascribing functions to the objects it encounters and thereby integrating them into a coherent system of its own" (1992: 42). But coherence to the human animal is always of a particular sort, shaped by particular symbolic systems. Our conception of the world is in some sense given, although our experiential knowledge is likely to modify our understanding of it. The human being is not only constituted *environmentally* (as Ingold argues) but also *socially* and *culturally*.

Ingold's argument revolves around his concept of affordances—what do things in the environment offer (provide or furnish) to an animal for good or ill? He believes that "our immediate perception of the environment is in terms of what it affords for the pursuit of the action in which we are currently engaged" (ibid.: 44). The difference between human ways of relating to the environment and that of other animals, however, is that the animal is limited by its genetic endowment—there are only a limited number of affordances that a given non-human species will find in its environment, and those affordances, with few exceptions, are hard-wired into its genetic constitution. What liberates us from our genetic endowment is culture. The conceptual scheme with which human beings operate will direct them to affordances that are beyond the reach of a non-human animal; and once those affordances are grasped, they will cease to be affordances in Ingold's sense and become symbolically encoded as having certain kinds of use and other value to particular communities of human beings. One problem with Ingold's position is that there has already been too much human activity—and concomitantly a great deal of ordering of a great many cultural projects—to leave many neutral uncoded objects in the environment (and notice that Ingold's concept of environment refers essentially to a bio-physical one). Even granting Ingold's point that culturally unmediated engagement with the environment is possible—about which I am skeptical—once it has taken place, as I have suggested above, it will be symbolically coded and become part of the cultural knowledge of the community. The salient point is not simply that culture allows human beings to shape their responses to the environment (or, as Ingold prefers, allows human beings to adapt the environment to their own cultural project, which is much the same thing, and indistinguishable in practice) but that culture also interferes with the ability of human beings to adapt—that is why people go hungry in an environment that they do not see symbolically coded for food (Redman 1999: 194).

The man who engages the environment by hunting for game is doing so according to certain modes of constructing his relationship to it that determines that certain animals are edible and others are not, that certain technologies are appropriate and others are not and that certain strategies will lead to successful outcomes and others will not, and so on. Richard Lee discovered that !Kung hunter–gatherers in the Kalahari did not utilize all of the foods that were afforded them by their environment, even in a drought year, and even though they were perfectly cognizant of their edibility; rather, they utilized the environment according to a conceptual scheme of their own (Lee 1969). As Bernard Cohn has pointed out, "There can be no practical realities without the symbolic coding of them as practical" (Cohn 1990: 41). Our engagement with the environment subjects our categories to a continuous interrogation and it is through the process of experience and reflection on those experiences that culture changes; but what we see as meaningful in our environment at any given moment is organized by our culture. We cannot *see* a reality that is not already organized by our categories. We experience reality but we understand it and give meaning to our experience in terms of the categories within which we operate.

Themes and concerns of the book

A number of themes link the chapters in this book. Not every theme surfaces in every chapter, but in what follows, I draw attention to some of the more significant ideas that are discussed in this book and how individual chapters are located in relation to these ideas.

Scientific discourse and local knowledge

The first theme, addressed in the chapters by Metz, Russell, Nightingale and Cameron, is the relationship between scientific discourses of the environment and local knowledge. Metz's Chapter 1 is an analysis of the discourse of crisis narratives, in this case, the Theory of Himalayan Environmental Degradation (THED), which he situates in a Euroamerican worldview that has come to dominate the global discourse of environmentalism. Malthusian ideas have become the common sense of Western perspectives on human–environment relations, especially when they are applied to the relations that poor people have with the environments in which they live. While the argument that poverty causes environmental stress is widely held, especially among policy-makers (Brundtland 1987; Beckerman, 1992; Barrett 1996), it has been critiqued on empirical grounds (see, for instance, Guha and Martinez-Alier's (1997) argument for an "environmentalism of the poor," and Moseley's (2005) examination of sustainable farming practices among rich and poor cotton farmers in Mali). Metz's interest in THED, however, has to do with its status as a particular kind of perspective, a symbolic system through which certain groups of people (most importantly, those involved in Nepal's development industry) have given meaning to a concept of environment.

The fulcrum of this worldview are two contrasting views of "indigenous people" or people who are part of "traditional cultures." One view treats them as "ecologically noble savages" who live in harmony with nature and who are used in the discourse of crisis to critique the environmental excesses of modern industrial society—the classic example in contemporary environmental discourse being the speech given by Chief Seattle in 1854, as imagined by a Hollywood screenwriter in 1972. The other worldview, which is explored by Metz, is a view of those same people that is darkly colored by the pessimism of Malthus, which has become common sense in some quarters; it is a view that disparages local knowledge and holds the alleged ignorance of the poor (not to mention their fecundity) responsible for environmental degradation—the ecologically ignoble savage, if you will.

Metz's chapter is an examination of a particular kind of understanding of the environment, one produced initially, in its modern form, in the belly of the West and then transmitted around the world where it has been taken up mainly by social and political elites. This understanding of environment sees it as fragile and vulnerable to human misuse. In Nepal, this view is held by policy-makers and development planners both foreign and local. THED has endured in these circles despite the sustained scientific critique to which it has been subjected, because, as Metz argues, it is a necessary aspect of their discourse; it justifies their presence and their engagement in the lives of ordinary people, and it provides ideological justification for the state to extend its control into areas that were once the prerogative of the local. It is a discourse that appeals simultaneously to "common sense" and to self-interest, and is therefore powerful. It also has the virtue of being "scientific," although as Metz points out in his chapter, the science is fundamentally flawed. But the scientific aura provides THED with both status and legitimacy. The conclusion that researchers have since come to, that THED is greatly exaggerated, would suggest that local people have not been mismanaging their environment after all, a conclusion much less appealing and useful, for instance, to Nepal's political elites and its advanced development industry.

But Metz intends more than simply to critique the way development planners and policy-makers understand the concept of environment; his chapter is an argument, drawing on critical realism, for the preeminence of a particular kind of knowledge, that produced through scientific inquiry. Even though the crisis narrative is based on scientific assumptions that are demonstrably misguided, the scientific method has built into it the ability to correct its own misuse or misapplication, which in fact is what happened with THED. Science thus promises us, Metz argues, a clearer understanding of reality than do other forms of knowledge.

Local knowledge tends to be valorized in the literature on development (see Hobart, 1993; Scott, 1998), a response to the devalorization of it in development discourse. There is a reality that exists independently of the way we know it (and we know it only in the terms in which it is filtered and organized by our categories), but the vast majority of human beings (except for the philosophically inclined, who are in a minority) take for granted that their understanding of reality is in fact reality as it actually is. Actual biophysical things and processes exist independently of the way we know them, but the way we know them may in fact affect the way they are. What is significant about THED is not that it was mistaken in its analysis of the Himalayan environment but that it emerged from an epistemology that has the capacity for self-correction. Does local knowledge of whatever sort also share that capacity for self-correction and, if so, in what sense and to what extent? How are local knowledge as well as trans-global forms of knowledge informed, shaped and constrained by structures of power, control and self-interest at every level?

The 2004 Asian tsunami illustrates this general point. Modern science understands it as the outcome of tectonic forces, which cannot be controlled but whose devastating impact for human society can be minimized by the implementation of a warning system based on modern technologies. Other ways of knowing might understand tsunamis differently. In Sri Lanka, for example, tsunamis have been few and far between. A possible early reference to one comes from the sixth-century AD Sinhala chronicle, the *Mahavaṃsa*, which records the actions of a second-century BCE king, Tissa, during whose reign, in the chronicler's words, "the sea overflow[ed] the land" (Mahanāāma, *et al.* 1986: 147). Tissa believed the gods were angry because he had put to death an arahant (someone who has achieved nirvana) on baseless suspicion of having an affair with the queen, and he made restitution by putting his daughter in a boat and casting her adrift in the ocean (ibid.). Tissa presumably made a culturally meaningful interpretation of the situation and acted in a culturally appropriate way but there is no question that the conceptual schemes by which scientists operate are likely to have a more positive impact on human well-being than Tissa's. While neither understanding of the tsunami can prevent it from happening, the first has the potential to save lives. In both cases, certain conceptual schema are used to interpret and act upon nature, but with very different likely outcomes.

The question of how science relates to local knowledge and to systems of power is taken up by Mary Cameron, Andrea Nightingale, and Andrew Russell. The most important thing about local knowledge is that it is in fact local, and very often practical, the product of generations of engagement with nature to transform it for social uses. Ayurveda, for instance, is the distillation of generations of practical knowledge of the environment (and Ayurveda's practicality and effect upon nature are recognized by pharmaceutical companies that see ethno-botanical knowledge as a potent source of future profit) coupled to systematic conceptions of how nature works that western science finds unfamiliar. In Chapter 2, Cameron locates such indigenous knowledge in the context of the power of the development imagination in Nepal, where development and progress are equated with scientific rationality and the paraphernalia of the modern state, including the possession of university diplomas. All of this serves to disadvantage and marginalize the traditional practitioners of Ayurveda, even as the value of their knowledge is recognized. Ironically, even as the practitioners are marginalized, their knowledge is valued, not because it is local but because it is seen as natural, and the quality of being natural has achieved a significant cachet today.

In Chapter 4, Nightingale deals with two models of knowing the forest: those of rural people in the Mugu district of Nepal, whose knowledge of the forest and its meaning emerges from their situated practice of making a living from it, and that of geographers, who draw on techniques of aerial photo interpretation and an inventory of the vegetation to develop an understanding of forest. For village people, divided along lines of caste and gender, the meaning of forest is embedded in usufruct rights; how the concept of forest is construed is a function of the relations that one has with other human beings, who may be in the position to facilitate or deny access to what an individual or family needs to survive or be a full citizen of the community. The geographer, on the other hand, brings to the idea of forest a well-established and defined concept in her science, which she then attempts to map on to the object of her inquiry—with mixed results. As Nightingale notes, "[the] boundaries, qualities and meaning [of forest] shift with each method used to explore it and from different perspectives, requiring one to question the existence of *a* forest" (p. 98).

In Chapter 3, Russell deals with two diverging understandings of the Himalayan environment, to call into question the tendency to romanticize the relationship that small-scale societies have with their environment. One—THED—is held by development planners who see degradation as the defining characteristic of the Himalayan environment. The other is the worldview of the Yakkha, a Tibeto-Burman ethnic group that lives in the hills of east Nepal. The Yakkha do not see the same environment that development planners do. For instance, the question of receding forest cover that preoccupies scientists and planners is not of especial interest to the Yakkha, perhaps because in their own locality, forests are not yet a scarce resource. The Yakkha instead have an ambivalent relationship to the forest; as the abode of predatory spirits and wild animals, it is at once a source of danger and fear, even as it is also the source of resources necessary for the survival of the household. This contrasts with Fortier's description of the Raute in Chapter 5, who see the forest as the locus of domesticity. Not surprisingly, any conservationist tendencies born of a recognition of the dependence on forest resources are counterbalanced by the desire to minimize the dangers inherent in the jungle itself by pushing back its frontiers. Perhaps, like the Greek peasants described by Theodossopoulos in his account of struggles over conservation on a Greek island (2003), what the Yakkha value and see as environment or nature is not the forest, a domain that represents forces outside their control, but the land that has been domesticated by human action.

All human societies have practical knowledge of the environment, without which they could not reproduce themselves. All such practical knowledge, whether characterized by scientific understanding as exemplified by THED, by the particular situated knowledge of Mugu villagers which allows them to negotiate access to resources, or of Ayurvedic practitioners, shifts and transforms with experience and their engagement with their reality as they conceive it.

Culture and social difference

A second theme that runs through this book is that how people conceptualize their environment is shaped, despite shared understandings with other members of their social group, by divisions of gender, social status, age, ethnicity and other social criteria. The individual's position in the social system as well as his experience shapes the way he understands the world. These issues are drawn out in the chapters by Nightingale, Russell, Subba, Campbell, Fortier and Aggarwal. For the people of Mugu described in Nightingale's Chapter 4, the concept of forest is mediated by the usufructary rights in forest produce that different segments of the village population enjoy. Men and women have different relationships with the forest, as do high caste and low caste people, and the nature of their relationship to other people shapes their understanding of what forest means. As she observes, the idea of forest refers not merely to trees but to a "host of political, social and material practices that are not necessarily visual" (p. 95). This is an important point from the perspective of policy-makers; as Nightingale points out, the failure to incorporate local elites has led to the failure of community forestry programs.

The contrast between the way settled agricultural societies and foragers like the Raute see the forest is starkly drawn by Fortier; she includes in her analysis the worldview of state bureaucrats, who are from agricultural or urban backgrounds and with whom the Raute must maintain good relations in order to carry on their way of life without impediment. The cultural scale she analyzes is that of the ethnic group (see also Subba's chapter); her chapter does not deal with the heterogeneity within the group but rather with the broad consensus reached about certain matters by its members. Like the Nayaka of South India, described by Nurit Bird-David (1990), the Raute conceive of the forest as a giving environment, and the model for their relationship to the forest and other creatures that dwell there is based on kinship. While farming people may have a sense of stewardship or caretaking where forest resources are concerned, the forest is primarily a place of danger, and to be feared (see Russell, Chapter 3 in this volume).

In Chapter 6, Subba paints a broad portrait of Limbu life, examining how Limbu identity is shaped by its environmental context. The cognized environment in which the Limbu—like the Yakkha—live is linked to their social and spiritual worlds and cannot be understood apart from this, but it does not follow that there is one monolithic Limbu way to understand the environment. Rather, these understandings vary across lines of gender and are subject to the transformations wrought by the external socio-economic environment in which Limbu society is embedded and to which it must respond. Subba also makes the important point that a cognized environment requires a sustained, practical engagement with the environment; when the pattern and intensity of that engagement change, cognition changes also.

Safia Aggarwal's Chapter 7 deals with villagers in Kumaun who conceive of nature in terms of its religious and ritual significance, and the consequences those beliefs have for natural resource use. Certain species of plants and animals, as well as geographic features such as forests and rivers, are accorded a special significance or are venerated because of their association with the divine. Sacred forests, a phenomenon much examined in the literature, is an example of this, and one that has been much idealized, as Freeman (1999: 257) points out, in the study of India's environmental history. Aggarwal focuses on two different kinds of forests associated with deities: temple forests, which belong to the deity and must be respected for that reason (see also Freeman 1999: 263) and communally owned forest-land (*pañcāyat* forest) that has been placed under the protection of a deity for a limited period of time in order to conserve it in the face of unsustainable use. She examines this question by showing that the efficacy of the strategy of placing forests under the protection of a deity not only varies across the population by social class and gender but it also depends ultimately on the availability of similar resources close at hand to which people may turn—in effect, environmental degradation is simply shifted to adjacent patches of forest. Although sacred groves have been idealized as an indigenous form of resource conservation, Aggarwal's chapter demonstrates the limits of this strategy, not to mention its social inequity. In Chapter 8, Mawdsley also discusses a variety of Hindu views on the environment, not all of which are conducive to an environmentalist ethic (and her chapter should be read together with Aggarwal's, who discusses how contemporary Hindu religiosity, with its emphasis on the built environment of temple complexes, undermines what protection is afforded the environment by sacred groves).

The synthesized environment

A third theme, discussed principally by Subba and Campbell is that the concept of environment does not readily translate across cultures. A coherent narrative of its meaning cannot be found in other places but must be synthesized. Indeed, most of the chapters in this book do not talk of the environment *per se* but have focused their discussion on the way human beings relate to, make use of, and process conceptually some tangible aspect of the physical landscape, such as forests and rivers. The concept of environment, despite its current international resonance, derives initially from the Euroamerican experience, especially the connotation that the environment is something that can be harmed and must be protected from human activity. Both Subba and Campbell show how we might understand what a concept of environment might look like in other cultural contexts, and the limitations of the concept for conservation practice. In Chapter 10, Campbell argues that not all domains of culture are equally coherent, and a concept like the environment has much less coherence in the Tamang way of life than, say, a notion like kinship, about which it is easier to discover a consistent narrative. Environment is a concept that we impose on the lived world of others because it is of significance to us and we wish, in essence, to see if others share our concerns. We frequently discover that they do not. Yet environmentalism is today a trans-cultural discourse (cf. Milton 1996) and we may reasonably expect that these systems of ideas will filter down and engage with conceptual schemas that shape the lived experience of people in villages all over the Himalaya. Stacy Pigg (1993) shows how this has happened with another transcultural discourse, that of developmentalism (or *bikās*, as it is known in Nepal).

Campbell's chapter critiques the over-determined idea of culture utilized by policy-makers, who assume it is characterized by a core cultural disposition and evenly distributed cultural knowledge. Such a reified view of culture treats it as one among a series of domains, and in the hands of policy-makers, one that lends itself to instrumental uses. More than that, however, Campbell seeks to displace the concept of culture as the analytical frame through which the way that human beings locate themselves and make their lives is to be analyzed. He draws instead on Ingold's idea of affordances to explore the forms of Tamang engagements with the world around them. The idea that environment stands apart from the human world is not, however, a universal idea even in the West, as Theodossopoulos' work suggests (2003), and conceptual schemes and practical engagement with the world in order to reproduce the conditions of one's existence are not mutually exclusive. As Campbell indicates, the Tamang do have a consensus at some level based on ideas about "the society of people, gods, demons and animals" (p. 200) that is being challenged and transformed by the intrusion of new ways of being originating outside these hill villages.

The environment as social critique

Activists can use the concept of environment to fashion critiques of society and polity, as Rademacher and Mawdsley show. Both deal with the uses to which rivers can be put in political and environmental discourse. Rivers, for Hindus, are sacred; the anthropologist Kelly Alley writes of the Ganga, "She is a mother, goddess, purifier, and sustainer of all life" (2002: 17). Yet the Ganga suffers from extreme pollution, as does the Bagmati, which flows through Kathmandu and has for Hindus similar symbolic significance. For environmental activists in Kathmandu, ecological degradation is a metaphor for cultural degradation.

Focusing on one prominent activist on the issue of the Bagmati's future, in Chapter 9, Rademacher discusses the importance of place, how places are imagined and how ecological degradation becomes a metaphor for the degradation of both polity and culture. The present condition of the Bagmati, which resembles nothing so much as an open sewer running through the heart of Kathmandu, evokes nostalgia for an imagined past among citizens of the valley who see the river as mirroring the condition of the decaying polity. Here too the way the environment (in this case, that of the river, which is linked to a semantic edifice that includes "the myth, pre-history, and history of an interconnected riverscape and Valley polity" (p. 170) is conceptualized, is grounded in a particular class and social status and thus in a particular social experience. The notion of a Bagmati civilization promoted by river activists would, were it to be realized, lead to the forced eviction of squatter communities that have colonized the old riverbed. The restoration of the dying river to its former glory is the restoration also of Kathmandu's imagined past and of an era of cultural stability and order. The condition of the river is a metaphor for the condition of the state and an expression of the unease of people with the existing society, which finds expression in a moral discourse about the Bagmati (cf. Grove-White 1993).

Rivers have their political uses across the border as well, as Emma Mawdsley shows in her discussion of the Vishva Hindu Parishad's intervention in the politics of the Tehri Dam. What are the consequences, Mawdsley asks, for environmental understanding and action when environmental issues are co-opted by politics? While the relationship between values promoted by various schools of thought within Hinduism and an environmentalist ethic are complex and problematic, as she points out, the problems are compounded when environmentalism is linked to nationalism. In the case of the Tehri Dam, the rhetoric used by the VHP (the World Hindu Council, a militant nationalist organization) to protest the dam's construction is linked to anti-Muslim actions, including the destruction of the Babri mosque in Ayodhya. The threat to the river from the dam is semantically linked to the threat to a "Hindu" nation from the Muslims in its midst; the Ganga, like the Bagmati, becomes a metaphor for the state of the state.

Conclusion

The essays in this volume call for a critical attitude to both the production of knowledge (in this case, about the environment) and the valorization of that knowledge. While scientific and technocratic knowledge is valorized by policy-makers, local knowledge and ways of understanding are just as universally dismissed—and policy-making oftentimes proceeds in ignorance of their very existence. Contrary to the romanticism of environmentalists like Vandana Shiva, local knowledge does not always conform to the category of conservation as understood by environmentalists and development planners, but it is always relevant, and whether or not local understandings of and approaches to environmental questions meet the ends of conservation, they must be taken account of. Development planners, meanwhile, tend to assume that their ends are transparent, meaningful and worthwhile, and seldom question the wisdom or assumptions implicit in the way they construe problems demanding their intervention. The people they seek to help operate in their own conceptual universes, however, as well as in their own social and political worlds, and much of the responsibility for the failure of schemes of "development" and conservation must be attributed to the inability or reluctance of planners to come to terms with this basic fact. Just as local knowledge must be examined case by case (Tissa was no guide on how to prepare for a tsunami), so too must the knowledge generated by science and the assumptions of global development and conservation organizations and the national elites who are linked to them; the particular cultural model of the environment exemplified by THED has had a profound impact on Himalayan peoples and has shaped a great many mostly failed or irrelevant interventions to improve the condition of their lives. All these forms of knowledge are also forms of power, but the power of technocratic and scientific knowledge is greater than the others and is a factor that shapes the course of their development even as it is relatively immune from reciprocal influence.

Note

1 The few exceptions include Gurung (1982), Ingles (1997), and Maag (1997).

1 Downward spiral?

Interrogating narratives of environmental change in the Himalaya

John J. Metz

Introduction

During the 1970s the notion that human activities in the Himalaya were producing an unfolding environmental crisis became an unquestioned discourse of the entire academic and development communities. The argument was most effectively presented by Eric Eckholm in an article in *Science* (1975) and a book (Eckholm 1976). Here is Eckholm's synopsis:

> Most know the kingdom [of Nepal] as the photogenic home of Mount Everest, as an exotic Shangri-la sprinkled with pagodas and quaint villages tucked away in the folds of the Himalayas. The facade of romance and beauty remains intact, but behind it are the makings of a great human tragedy. Population growth in the context of a traditional agrarian technology is forcing farmers onto ever steeper slopes, slopes unfit for sustained farming even with the astonishingly elaborate terracing practiced there. Meanwhile, villagers must roam further and further from their houses to gather fodder and firewood, thus . . . ground holding trees are disappearing fast . . . Landslides destroy lives, homes, and crops . . . Topsoil . . . is now Nepal's most precious export, but one for which it receives no compensation.
>
> (Eckholm 1975: 189–90)

Eckholm and others alleged that the processes were a series of interlocking vicious circles. After World War II, populations of subsistence farmers exploded, forcing families to clear forests to feed their new members. With fewer forests, fuelwood became scarce, making women devote increasing amounts of time to collecting it. Fuelwood shortages led people to burn dung rather than fertilize their fields, which intensified food shortages and pushed desperate farmers to clear forests on even steeper slopes. The intense monsoon rains now ran off the deforested slopes, creating floods, accelerating erosion in the uplands during the monsoon, and drying up springs in the dry season. As the runoff and sediment from the mountains poured onto the rivers, plains, and deltas of the Indian and Bangladeshi lowlands, the lives and livelihoods of hundreds of millions were threatened (e.g., Sterling 1976; Ives 1980; Bowonder 1982; Claiborne 1983; WRI 1985; Myers 1986).

Though the scenario that Eckholm articulated had been argued by environmentalists and forestry bureaucrats for a century, his presentation seemed so intuitively obvious, was so felicitously written, and came at a time of growing environmental concern that by 1980 it was universally accepted by scholars, foreign aid donors, and national planners. The power of the narrative is illustrated by its acceptance by writers from opposite poles of the ideological spectrum: it became the basis of a *Harpers* article by Claire Sterling (1976), who also promoted US military interventions around the world (Sterling 1981), and of Blaikie, Cameron, and Seddon's (1980) neo-Marxist critique of Nepal's development.

The numerous aid projects operating within Nepal during the late 1970s and 1980s all began to include components designed to address this environmental crisis, and some projects (e.g., USAID's Resource Utilization and Conservation Project, Nepal Australia Community Forestry Project) focused exclusively on forests and environment. In addition, foundations and research organizations began to fund scholars seeking to specify the degradation processes. As academic researchers undertook studies of how communities were using their forests, agricultural fields, and wild lands (Donovan 1981; Wallace 1981; Johnson *et al.* 1982; Bajracharya 1983a, 1983b; Fox 1984; Mahat *et al.* 1986a, 1986b, 1987a, 1987b; Byers 1987a, 1987b; Metz 1990; Zurick 1990; Brower 1991; Stevens 1993), some development project workers did revealing research into various aspects of the hypothesized problems (e.g., Levenson 1979; Wyatt-Smith 1982; Carson 1985; Nield 1985; Gilmour *et al.* 1987; Gilmour 1988). Of the academic researchers, Jack Ives and Bruno Messerli received a grant from the United Nations University to study highland–lowland interactions, and this funded a series of studies at high (Byers 1987a, 1987b) and mid-elevation sites (Johnson *et al.* 1982). In 1981, Ives began a journal, *Mountain Research and Development*, which became a major outlet for this research.

Much of this work, however, found the assumptions of the crisis narrative inaccurate. As these results accumulated, Ives and associates organized a conference to review the issue in April 1986 at a resort in Mohonk, New York, and invited scholars and practitioners to share their findings. The papers presented challenged the crisis formulation, which Ives (1987) labeled the 'Theory of Himalayan Environmental Degradation' (THED), and led most of the participants to reject THED as at best vastly over-simple and at worst as simply wrong. *Mountain Research and Development* (Ives *et al.* 1987) published many of the papers and concluding documents from the Mohonk Conference, and a new academic consensus began to emerge challenging the crisis discourse. Ives and Bruno Messerli summarized the THED and many of the arguments against it in a 1989 book, *The Himalayan Dilemma: Reconciling Development and Conservation*. Ives recently summarized research completed since 1989 to further confirm the revisionist position (2004).

Nevertheless, at the Mohonk Conference, scholars from India maintained that the crisis was real and serious. Moreover, scholars and policy-makers from India, China, and Thailand continue to dismiss the international consensus and promote the crisis narrative (Tiwari 2000), which has prompted Ives (2004) and Blaikie and Muldavin (2004) to investigate why this disjuncture exists.

In this chapter my primary goal is not to evaluate the truth or falsity of these claims: much evidence outlined below shows that forest degradation is occurring, but its intensity varies greatly by location, as Chapters 3 and 6 in this book by Russell and Subba illustrate. Rather, I hope to consider these changing narratives of human–environmental interactions in order to explore the ideological, institutional, political, economic, and psychological factors that give the narrative its power. These discourses form a part of the culture of scientists and development professionals from both the donor and recipient countries. As Guneratne argues in the Introduction to this book, although culture is the dynamic consensus a group of people share to represent their social experience, the contours of that consensus will vary across the subgroups of that culture. Hence, I want to explore the reasons why the subgroups of this developmentalist culture have or have not interrogated and rejected THED. I also want to consider the epistemological and ontological bases which allow us to evaluate the validity of the competing claims, and finally I hope to review the ways scholars and practitioners can improve our understanding of human–environmental interactions in the Himalaya.

The first section summarizes the fluctuating discourses on the impacts of human beings on the Himalayan environment. The second section examines the unsupported assumptions of the crisis narrative, borrowing from analyses of similar crisis narratives about the African environment. The third section reviews the multiple factors that have made scholars and policy-makers embrace crisis narratives. The chapter's fourth part reviews suggestions to allow scholars and practitioners to conceptualize, research, and manage human–environmental interactions more effectively. The fifth section identifies the cultural and epistemological lenses this chapter uses to view the Himalaya and the world beyond. The conclusion summarizes the chapter.

The Theory of Himalayan Environmental Degradation (THED) and its critics

The environmental crisis narrative of the Himalaya that gained widespread acceptance in the 1970s had a long history. An Indian forester articulated the argument in 1960 (Kaith 1960), as had a UN forester six years earlier (Robbe 1954), but the argument had been used by British colonial foresters in the late nineteenth century to justify "reserving" forests for use by the colonial state, thereby alienating local forest users from the fodder, litter, timber, foods, medicines, etc. that they had traditionally collected and needed for subsistence (Gadgil and Guha 1992; Hobley 1996; Saberwal 2000). Indeed, the use of forests became a struggle between the colonial land revenue administration, which encouraged forest conversion to taxable agricultural lands, as had the native kings of the pre-colonial period (Rangan 2000a) and the Forest Department, which wanted to use the timber for railroads and other infrastructure. The Forest Department won this contest and then transferred the environmental degradation arguments developed in India to their African colonies (Leach and Mearns 1996b).

The dominance of the degradation discourse in the 1970s began at a conference on mountain environments in 1974 sponsored by UNESCO and the German development agency, GTZ. Eckholm was the *New York Times* science editor at the time and had been invited to the conference by the organizers (Ives 2004: 2–3 and 212–13). His advocacy of the narrative via an article in *Science* (Eckholm 1975) and a book (Eckholm 1976) contributed to its popularity, but clearly other forces were also promoting the position.

Some of the first challenges came in the early 1980s. Students in Ives and Messerli's project (Johnson *et al.* 1982) realized that mountain farmers were skilled interpreters and manipulators of the landscape, assiduously maintaining agricultural terraces, identifying when terraces would inevitably fail, and using irrigation water to reshape landslides to re-create them. John Wyatt-Smith (1982), completing his long career as a forester in tropical Asia, assembled the few good studies of forest use that were available to argue that provisioning livestock was the largest demand on forests in Nepal, dwarfing fuelwood and agricultural field creation. An undergraduate on a year abroad program, Deanna Donovan (1981) collected the 49 estimates of fuelwood consumption that had been reported to find that the largest was 67 times the smallest; she concluded that this was due to poor or non-existent research methods. Fox (1984), Wiart (1983), and Metz (1990) demonstrated what effective methods of measuring forest product use would be: year-long residency and the actual weighing of products used by a representative sample of households. Byers (1987a), another of Ives' students, showed that erosion near Mt. Everest was significantly less than had been hypothesized and that forests in Khumbu had actually increased slightly since 1952 (Byers 1987b), rather than declining precipitously as others had alleged (Fürer-Haimendorf 1975). Carson (1985) and Ramsay (1986) argued that mass wasting rather than overland flow produces the vast majority of erosion and sediment and that continual uplift of the mountains made mass wasting inevitable, as it has for millennia. Hamilton (1987) distinguished between forest conversion to other cover types and forest degradation, a distinction that is absent from the crisis narrative. He further noted that no studies existed which estimated how long sediment eroded in the mountains takes to reach the lowlands, suggesting that it could take scores of years for today's mass wasting sediment to reach India. Gilmour *et al.* (1987) measured infiltration capacity of soils with varying vegetative covers and found very few precipitation events exceeded the infiltration capacity of even the most degraded sites. Further, Gilmour (1988) found massive plantings of trees on private lands in deforested lower elevations of the mountains. Nield (1985) compared air photos from 1964/65 to those from 1978/79 to discover that the area with forests in the uplands had remained constant, though many of those forests had been degraded and that forests of the lowland Tarai and Inner Tarai were rapidly being cut. Mahat *et al.* (1986a, 1986b) used the comprehensive historical research of Mahesh Regmi (1971; 1978) and Mahat's own investigations to show that indigenous kings of the eighteenth and nineteenth centuries probably doubled forest conversion to agricultural lands through their taxation policies, so forest conversion was not simply due to the subsistence needs of yeoman farmers (see also Rangan 2000a). Messerli and Hofer (1995) have argued that flooding in the lowlands is caused more by precipitation and land use practices in the lowlands than in the mountains (see also Alford 1992). These and subsequent studies (see Ives 2004, for a comprehensive review) challenged the crisis narrative and led to its rejection by most international scholars as vastly exaggerated and deceptively over-simple. The consensus is that although there are places where human actions are producing environmental degradation which undermines subsistence, the Himalaya is so tremendously diverse that generalizations are questionable (Blaikie and Muldavin 2004).

Unsupported or erroneous assumption underlying the crisis narratives

The crisis narrative of the Himalaya and other regions of the world (see Leach and Mearns 1996a, 1996b) are based on assumptions about societies, science, and the environment that are without support, are over-simple, or are erroneous. These assumptions can be collapsed into four.

The first is the belief that although indigenous people are the agents and victims of their ignorant use of the environment, they are not capable of changing their practices to adjust to the social and environmental changes that their activities produce. Implicit in this is the belief that their indigenous knowledge about their environments can provide no help in creating more sustainable systems of use (Gilmour and Fisher, 1991; Gilmour and Nurse, 1991). Neo-Malthusian assumptions share this denigration of local ability by viewing local people's reproductive behavior as an independent variable, beyond the control of conscious decision, rather than as a variable dependent upon the economic, social, and personal goals of people (Nag *et al.* 1978; Cain 1978; Fricke 1986; Guha 1990; Leach and Mearns 1996b; Hoben, 1996).

The second set of assumptions holds that current, "snapshot" perceptions of contemporary land-use practices can be extrapolated to larger historical and spatial scales. Hence, the crisis narrative completely fails to examine the historical and spatial patterns of population growth or collapse, land use patterns, indigenous and colonial state actions, geological processes, etc. (Leach and Mearns 1996a; Mahat *et al.* 1986a, 1986b).

The third set of assumptions reflects the faith in the ability both of western science and western development agents to objectively identify the problems of and solutions to unsustainable land use practices. This set includes assumptions about processes of deforestation, erosion, and indigenous peoples' use of their environments. For example, Stocking (1996) provides a cogent example in his discussion of the biases and errors inherent in "scientific" studies of erosion: (1) gullies are not the cause, but a symptom of a degraded catchment above the gully, so trying to stabilize the gully is the wrong solution; (2) surfaces with exposed stones do not necessarily show severe erosion, but may result from stones being brought to the surface by shrink and swell clays; moreover stones may protect soils by absorbing the impacts of raindrops; (3) vegetation in general does not prevent erosion, because its effects depend on the type of plant, its height above the ground, whether its leaves concentrate light rain into larger drops, and whether it channels precipitation down stems; (4) erosion rates extrapolated from soil-loss plots (bounded areas 3 by 20 m with a trough at the bottom to collect overland flow and eroded sediment) interfere with "natural" processes by blocking water and sediment flowing from above, by failing to record the soil moved and re-deposited within the plot, and by over-sampling clays and silts so that smaller plots record relatively larger soil losses than larger plots, thereby exaggerating erosion within large catchments by a factor of 100; and (5) valley deposition of eroded soil and nutrients creates the agricultural productivity of lowlands and floodplains, so lowland farmers may not see upland erosion as "bad." I discuss this in such detail because it is a powerful example of how "scientific" field methods can skew information provided to policy-makers.

The fourth set of assumptions results from the traditional ways geologists and ecologists have thought about environmental change. A bedrock assumption of geology is uniformitarianism, the idea that low magnitude, high frequency events and processes overwhelm the effects of high magnitude, rare events. The import to Himalayan landscape evolution of mass wasting events, triggered by extended tropical cyclone rainfalls and/or by earthquakes (Starkel 1972; Carson 1985; Bruijnzeel and Bremmer 1989; Gerrard and Gardner 2000, 2002) challenges the uniformitarian assumption. In ecology, the belief that ecosystems are usually in "equilibrium" with their climate and quickly return to equilibrium after a disturbance, articulated in Clements' concept of the "climatic climax" (Clements 1936) and embedded in the application of systems analysis to ecology (Odum 1969), long dominated ecological thinking and is still applied in the Himalaya, but has been thoroughly undermined by recent studies which show that disturbance is the norm not the exception (Oliver 1981; Pickett and White 1985; Christiansen 1989; Botkin 1990). Not only are most ecosystems so frequently disturbed by non-human and human forces that they never reach a stable state, but the "recovery" from a disturbance can produce significantly different assemblies of species, depending on the type of disturbance; for example, the forest that grows after a fire will have a very different species composition from the one that develops after a blowdown (Henry and Swan 1974). Yet, the equilibrium assumption continues to dominate thinking about Himalayan landscapes, thereby making any changes hypothesized to be coming from human subsistence activities "unnatural" and therefore catastrophic.

Epistemological challenges of THED as narrative

The foregoing discussion shows that the interpretation of human–environmental interactions in the Himalaya is just that: an interpretation, a "narrative." It is an example of the widespread recognition over the past 50 years that all knowledge is socially constructed. This contrasts with the belief that knowledge is the process of discovering some truth or reality, an essentialist process of reifying our concepts by assuming they correspond to and precisely identify entities existing beyond our consciousness. The realization that human knowledge is socially created poses a fundamental epistemological challenge to our ability to understand and effectively influence human impacts on the Himalayan environment.

At the simplest level it is obvious that knowledge exists in human minds and is transmitted from person to person, so it is a human creation. This realization underlies the understanding of culture that Guneratne describes in the Introduction to this book. As he explains, culture and discourses within cultures are webs of symbols that groups of people spin to engage each other in the process of sustaining and reproducing themselves and their society. The difficulty arises—and this is the central epistemological and ontological problem of philosophy—over the relationship between human knowledge and an external world: how do we, indeed, can we, accurately know a world that exists independent of our awareness? In its most extreme form, the belief that knowledge is a social creation has been interpreted to mean that knowledge only exists within the minds of individuals, that we cannot know an external reality. Yet, that humans can and do wrest a living from their environments shows that knowledge corresponds closely enough to an external world that this idealist assumption is untenable.

Nevertheless, that human knowledge is imperfect and a social construction challenges the realist position that we can know an external reality. Even in the hard sciences, Kuhn's (1962) studies of the history of physics show that physical theory is the creation of the society of physicists. Kuhn argued that science advances not in continuous small increments, but in abrupt revolutionary changes, as the community of scientists abandon one "paradigm" to accept another.

According to Kuhn, a mature science has a dominant paradigm that is accepted by all reputable scientists and which defines meaningful problems, acceptable types of methods, and expected results. Newtonian physics was the first such paradigm. Scholastic assumptions of the Middle Ages held that the heavens were a perfect realm, completely different from the imperfect earth. Copernicus argued for a heliocentric world and Kepler uncovered the mathematical regularities to predict its planetary movements. They challenged the earth-centered Ptolemaic cosmology, with its assumptions that only circular motion was sufficiently perfect to suit the heavens, and a series of crystalline spheres rotating around the earth creating the movements of the stars and planets. The Ptolemaic system could also predict where the planets would appear at a given time, but they were mathematically much more complex and less accurate than Kepler's predictions. Meanwhile, Galileo was identifying the laws regulating the movements of objects on earth, largely from his interest in exploring cannonball trajectories. Newton was able to integrate Kepler's laws with Galileo's investigations by deriving both sets of observations from one set of assumptions about gravity that affects all objects on earth and in the heavens. For the next 200 years Newton's synthesis was embraced by the community of scientists as defining what physics was about.

After a paradigm is accepted as meaningful, there is a period that Kuhn calls "normal science" during which the implications of the assumptions of the dominant paradigm are explored by scientists. However, experimental results occasionally emerge which contradict the expectations of the researcher and the assumptions of the paradigm. The scientist assumes he has made an error and performs the experiment again. If the result persists, he or she rejects it as impossible and seeks to find the error. As these anomalous results accumulate, however, scientists are puzzled, but generally explain away the disconcerting results. Eventually, some genius, invariably a young person, takes the anomalous results seriously, perceives that the assumptions of the dominant paradigm are unfounded and proposes a new paradigm based on new, more comprehensive assumptions, which explain the anomalous results and incorporate the old paradigm as a special case of the new model. Science enters a 'revolutionary' era where the fundamental assumptions of the science change. In physics, relativity and quantum mechanics are examples of this paradigm revolution.

In other sciences, like biology with evolution and geology with plate tectonics, the initial, defining paradigms have relatively recently become accepted and are still being elaborated. During the period before a single paradigm is accepted by the community of scientists, proponents of competing theories struggle for acceptance. This pre-paradigm period devotes considerable efforts to struggles over defining methods, the types of data one should collect, the goals of the discipline, etc. The social sciences appear to be in this pre-paradigm situation. Hence, even in the most developed sciences, the realization that knowledge is a social construction places human beings in an epistemological quandary that challenges our ability to understand and manage our world. I will return to this challenge in the fifth part of this chapter.

Explanations of the power and persistence of THED in development discourse

Several scholars have explored the Himalayan crisis narrative as a social creation and discursive enterprise and sought to explain its origin, reproduction, and persistence (Guthman 1997; Forsyth 1998; Blaikie and Muldavin 2004). These efforts are paralleled by scholars working on development and human–environmental interactions in Africa, and this African work complements Himalayan studies nicely (Roe 1991; Thomas and Middleton 1994; Leach and Mearns 1996a, 1996b). Below I combine these sources to explore why THED continues to be used to describe human impacts on Himalayan environments.

To solve problems, development planners make simplifying assumptions

Regarding the Himalayan environmental crisis narrative, Roe's discussion of development planning (1991) helps explain its power: development planners need "enabling assumptions," which are narratives that tell what will happen, so that they can overcome the profound uncertainties surrounding the problems they are charged to solve;

> The more uncertain things seem at the microlevel, the greater the tendency to see the scale of uncertainty at the macrolevel to be so enormous as to require broad explanatory narratives that can be operationalized into standard approaches with widespread applications. The unresolved failure of project blueprints derived from development narratives thus often serves only to reinforce, not lessen, the perceived need for some sort of narrative that accounts for the resulting increase in activity.
>
> (ibid.: 288)

Hoben (1996) suggests that when these narratives become actualized in "specific development programmes, projects, packages and methodologies of data collection and analysis," they become the 'cultural paradigm' associated with the narrative, and this paradigm comes to dominate the development institution.

Development planners make decisions on political not rational grounds

Despite the rejection of the crisis narrative by most international scholars and practitioners, it remains the unquestioned base of scholarship and land use policies in both India and China, and this induced Blaikie and Muldavin (2004) to explore why. They argue convincingly that development policy planners do not act according to the "rational model" (ibid.: 525), i.e., the belief that independent and objective scientists empirically analyze problems and propose solutions and that the policy-makers then use the scientific analysis to create a program to remediate the identified "problem," but rather that the policy-making process is "discursive" and political (Figure 1.1). Scientists are also affected by their disciplinary biases and assumptions and by the goals, assumptions, and directives of the funding agencies, with the result that competing scientific truths about the "problem" emerge. The

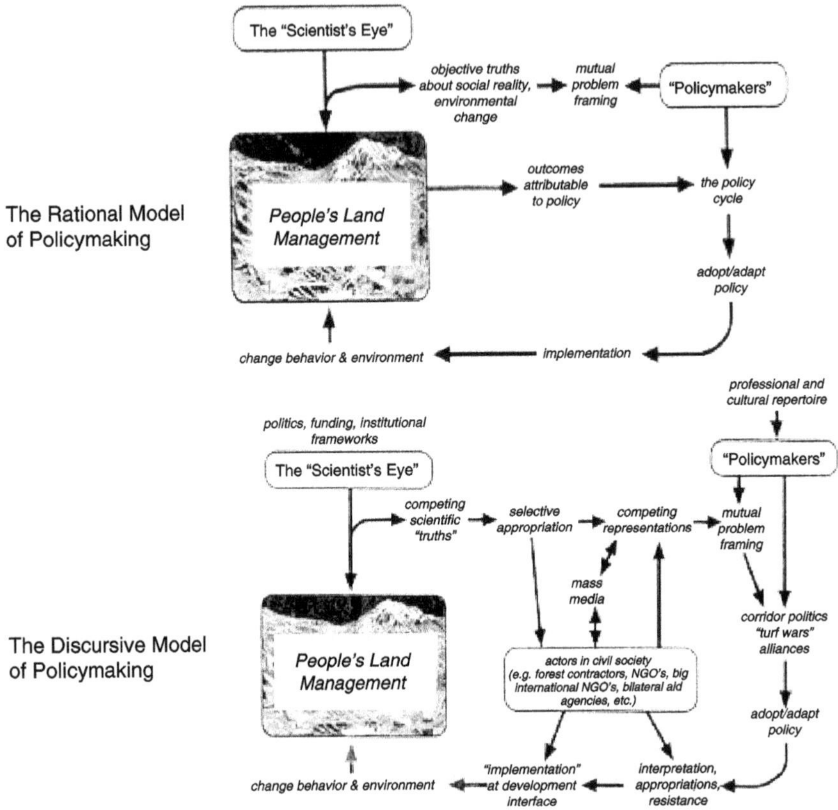

Figure 1.1 Models of policy-making.

policy-maker balances the claims and pressures of the many stakeholder groups who have conflicting interests regarding the policy being selected. The contestation takes place in competing discourses about the problem, and these are based on how the various groups believe they will benefit or suffer from what they expect the policy will create. Examples of these stakeholder groups include various economic classes living within the areas affected: powerful individuals within the government and civil society; various government bureaucracies; NGOs at all scales; donor governments and their development agencies; multilateral donors; and the mass media. The policy-maker then chooses the policy to balance these groups and then articulates the policy in government documents.

> The policy then finally appears at lower levels, often unannounced and unwelcome, and people at the local level make what they can of the "it" to which they have access. For example, local forest rangers or other government servants, such as village leaders, interact with local villagers

> in daily life, at "the development interface" of a particular policy . . . Then the policy, as it were, finally sinks into the sands of contingency, often unrecognizable as the outcome of an original set of intentions in the name of the public good.
>
> (ibid.: 526)

Given this chaotic process, it is hardly surprising that the policy-maker will choose policies which reflect the interests of the more powerful groups and coalitions of groups, and when those groups favor the crisis narrative, as in India and China, it will be adopted regardless of international scholarly consensus.

Crisis narratives makes outside experts stakeholders in land use decisions

Roe (1995) points out that claims by "expert elites" that local people are degrading their environments, especially when the results of the degradation are likely to affect groups beyond their territory, justifies the experts' claims that they are stakeholders in the struggle to control land use. Moreover, because of their expertise and credentialed knowledge, their interpretations should dominate those of others. "Whether right or wrong, the claims, counterclaims, and changing claims of experts serve principally to reinforce and widen the belief that what they, the experts, have to say really matters solely by virtue of their expertise" (ibid.: 1066). Knowledge justifies and reproduces social power (Foucault and Gordon 1980).

Changing discourses of human–environmental interactions adjust to changing narratives of development theory

These changing narratives operate at varying spatial scales. Forces supporting environmental crisis narratives can be viewed as operating within a nested hierarchy of scales. This is not to assume that scale is an a priori category of analysis, but rather is one that is continually reconstituted by struggles between "individuals and social groups in a perpetual socio-spatial power struggle" (Swyngedouw 1997: 141), much as was described by Blaikie and Muldavin (2004) for the national policy-making scale.

Recognizing that actors at a variety of spatial scales are affecting Himalayan environmental use patterns, I find the following scales of analysis useful: international, national, regional, local and personal. The actual scales that apply will be defined by each specific case.

At the international scale

During the colonial period, occupying administrations used claims of environmental degradation to justify taking control of resources from native land users. The "deforestation causing erosion" line was employed by British foresters in India to justify the "reserving" of forests for timber production, excluding indigenous users (Gadgil and Guha 1992; Saberwal 2000). This narrative was taken to Africa by British foresters, where it was transformed into overpopulation, deforestation, overgrazing, desertification narratives that persist into the twenty-first century (Leach and Mearns 1996b). The coercive power of the colonial state struck hard at villagers and so induced them to resist in every way they could (e.g., Scott 1985; Guha 1989; Gururani 2000). After the East India Company had defeated Nepali armies in 1816, an uneasy truce defined Anglo-Nepal relations until 1846, when Jang Bahadur Rana seized power from the Shah dynasty which had created Nepal in the last half of the eighteenth century. Jang Bahadur's assistance to the Company in the 1857 Sepoy Rebellion cemented the symbiotic relationship between the Raj and the Rana regime, through which the British assisted the Ranas in maintaining their despotic control by isolating Nepal from any modernizing influences in exchange for peace and the opportunity to recruit Gurkha soldiers (Regmi 1978).

In the postcolonial world, independent states generally continued the colonial state's policies. At the international level foreign aid gave donors the ability to impose their development theories on aid recipients, as the case of Nepal clearly illustrates. Julie Guthman (1997) links the changing environmental discourses in the Himalaya since 1950 to the changing models of development theory in the post-WWII period and to changes in Nepal's forest policy, as outlined by Hobley and Malla (1996).

Guthman identifies three aid regimes in Nepal: (1) Modernization (1950–73); (2) Basic Needs (1974–84); and (3) "Neo-liberalism" (1985–present). Each of these regimes had particular constellations of donors, economic sector emphases, and proportions of grants to loans, and each was reinforced by different sets of dominant ideas regarding the goals and strategies of development which had clear impacts on how to manage the environment (ibid.: 47).

The *Modernization* period began after the despotic Rana family was overthrown in 1950. The new government adopted "development" as its goal and began a series of five-year plans in the early 1950s. Modernization asserted that industrial-led development would create prosperity by encouraging local industrial development. Developing physical infrastructure could facilitate that process, so almost all aid was in the form of grants and devoted to infrastructure, especially roads and power generation. Many donor countries began projects, but only India and China provided significant amounts. Interest in the environment was minimal, although a UN forester invoked the degradation discourse that British colonial foresters had created in India (Robbe 1954).

The second period, called *Basic Needs*, sought to remedy the failure of modernization to trickle jobs and wealth down to the masses by intervening in ways that more effectively benefited the poor, while maintaining economic growth. This was the period when the crisis narrative took hold.

By 1980, most of the scores of bilateral, multilateral, and international NGO projects operating in Nepal had environmental components. In response to the pressures to conform to the basic needs rhetoric of political decentralization, popular participation in development activities, and collective control of resources, Nepal's Forest Ministry adopted a National Forestry Plan in 1976 and Community Forestry Act of 1980. These laws acknowledged the centrality of forests to rural subsistence systems and gave limited legal control over some forests to local political units, the Panchayats.

During these years academic and donor research recognized that local people were already managing forests to varying degrees and that government interventions had disrupted that management (Molnar 1981; Campbell and Bhattarai 1984; Gilmour and Fisher 1989). The populist message that local people shape their landscape and can be engaged in development fit well with Basic Needs rhetoric. Community forestry came to be seen as essential to improved use of forests and wild lands. Pressure by bilateral and multilateral donors induced the government to conform, though resistance in the bureaucracies to the loss of their power slowed the process.

The election of Reagan to the US Presidency, the global debt crises, and the failure of Basic Needs projects to solve agricultural problems ushered in the third, *Neo-liberal*, period. Neo-liberal ideology maintained that free markets and free trade would solve development problems and opposed all governmental attempts to intervene in the economy. During this period Community Forestry rhetoric became reality, as the Forestry Department devolved control over forests to local Forest User Groups (FUGs) (Gilmour *et al.* 1989; Gilmour and Fisher 1991). By 2002, 11,000 FUGs legally managed over 847,000 ha or 23 percent of the potential forested land in Nepal (Lachapelle *et al.* 2004: 2), though the vast majority of FUGs are controlled by elites and keep poor members powerless (Mahapahtra 2001; Malla 2001; Malla *et al.* 2003; Lachapelle *et al.* 2004).

Guthman (1997) argues that Neo-liberal ideology supported the expansion of Community Forestry; she also notes how the awareness of the discursive nature of human–environmental interpretations became widespread, though the crisis narrative remains embedded in the narratives of some environmental and development institutions (WRI 1995; Overseas Development Agency 1997) and in policies of many Himalayan states (Blaikie and Muldavin 2004). She provides a thought-provoking analysis, but one which fails to emphasize the ways that government technocrats and village forest users adjusted to, diverted, and shaped donor policies to their personal and institutional ends, as Blaikie and Muldavin illustrate. Moreover, because Community Forestry gives control to local forest user groups rather than to state bureaucracies, it appears to conform to Neo-Liberal philosophies. However, the control is given to *groups* who are expected to manage the resource effectively. This argument, and the entire discourse on "common property resources" (Ostrom 1990) are antithetical to the Neo-liberal solution of converting the commons to private property. I conclude that Guthman's argument that international forces shape development narratives and policies provides some insight, but must be supplemented by causal factors at smaller scales of resolution.

At the national scale

By including actors at the national scale we enhance our understanding of the power and persistence of the crisis discourse. First of all, the crisis narrative justifies extending bureaucratic power into lands and resources that had not previously been within their control. Second, adopting, promoting, and making policies that address the crisis narrative brought foreign funding to national ministries which allowed bureaucrats to expand their personal and institutional power and to siphon some of the funds into their own pockets. Guthman noted that donors demanded a new Ministry of Soil and Water Conservation be created, implying that this was an example of foreign domination of the development process, but this can also be seen as an opportunity for bureaucracies to expand. This position, of course, assumes that all national bureaucrats and agencies will prostitute their independence and integrity to foreign dictates in order to expand their personal and institutional power, but not all individuals will respond in this way.

Another perspective on the national scale influences are those of the donors. Why do they give aid? Part of the goals of all donor countries is to help the poor, as their rhetoric proclaims, but there is probably an inverse relationship between the importance of that goal and the importance of the country in the world political economy. Powerful countries use aid for political goals: to gain support in international arenas; to prevent revolutionary movements from creating instability and establishing a hostile new government; to facilitate the economic expansion of domestic corporations and political donors; to weaken the country's independence so they become more reliant on the donor. Larger, more powerful countries that receive aid are freer to chart their own course, and this partly explains why India and China can resist the change in environmental narrative.

At the regional and local scale

At the regional and local scale, individuals and groups will support, oppose, and subvert policies. With the emergence of policies designed to mitigate the environmental crisis, local people reacted in ways they interpreted would benefit them, as Nightingale's and Rademacher's chapters in this volume illustrate. In the early 1980s period of nurseries and plantations, local leaders and groups supported nurseries when the water system built to water the seedlings was also available for home consumption (Gilmour pers. comm. 1990). People planted trees when they were paid to do so, but they protected plantations only where they were convinced they would be able to keep some of the forest products. As community forestry laws changed to expand authority of user groups over the forests they manage, users engaged with the Forest Department to fulfill the bureaucratic requirements. Nevertheless, these communities remain highly stratified by caste and wealth, so the more powerful members of community forest user groups commonly take control of decision-making and are likely to garner more than their share of forest products (Malla 2001; Lachapelle *et al.* 2004). An example of a counter-current to the stratification of wealth and power at the local and regional level was the emergence and spread of the Maoist insurgency in Nepal.

At the personal scale

Individuals will respond to opportunities their positions afford to enhance their personal wealth and power. International development personnel earn high salaries and live more luxuriously than they can at home, with large houses, servants, and private schools for their children. US citizens do not have to pay US income taxes if they reside overseas for longer than a year, a 25 percent bonus on their salaries. Not uncommonly they do not understand the host country institutional barriers to accomplishing their development goals, like "corruption" and nepotism. By the time they do realize that they are unlikely to be able to succeed as they had hoped, they may have a relatively short time left on their contract, so they continue half-heartedly or they come to believe that they must be "realistic" about their situation and "work within the system."

In the same way host country bureaucrats and local leaders may exploit their position to extract the most money and power they can for themselves and their families. The widespread reality of "corruption" must be due to structural roots as well as personal mendacity (IDS 1983; Kondos, 1987; Metz 1995) because siphoning agency funds for personal aggrandizement is the norm rather than the exception. Hence, workers at all levels will promote the discourse that pays their salaries and provides opportunities to care for family and friends.

Yet, some expatriate experts and host country bureaucrats are able to overcome the impediments to make significant differences. The origin of community forestry in Nepal was due to the joint work of an expatriate Australian forester and a Nepalese District Forest Officer who was willing to risk bending the bureaucratic rules to give management of a local forest to village users, because they both had come to realize that without local cooperation management was impossible (Mahat pers. comm. 1985). The Nepal Australia Community Forestry Programme emerged from that collaboration as a pilot program for community forestry, and it has continued to field extraordinarily successful personnel.

Identifying causality along a "structure–agency continuum"

The above discussions identify both social structures and personal agency as causes of the fluctuating ideas, policies, and programs addressing human–environmental interactions in the Himalaya. Leach and Mearns (1996b) note that the contributors to their book on African environmental narratives (Leach and Mearns 1996a) vary in the degrees to which they parcel out responsibility between structure and personal agency in this issue in our post-structural period (Zimmerer 1994; Harré and Bhaskar 2005; Chowdhury and Turner 2006). Operating from a "post-positivist," but epistemologically realist position similar to what I describe below, Chowdhury and Turner offer a succinct summary of the dynamic between structure and agency at the scale of the local land manager, but the logic of their argument can be applied to agents at any scale:

> Agents do not make decisions independent of the political economic conditions in which they exist and the cultural and historical experiences that give rise to path-dependent constraints on their options. These conditions and experiences shape the very essence of household behavior, and together with biophysical considerations, they mediate the decisions made. Likewise, these conditions are not immutable, and the household, being composed of reflexive agents, interprets and reshapes its circumstances, requiring attention to household decision logic relative to the observed outcomes. In some cases, external forces appear to overwhelm the household and elevate the role of structure in understanding the outcome of the coupled system. In others, the household displays considerable latitude in decision making, and the outcome cannot be understood absent attention to the household behavior. Comprehensive understanding, according to this claim, requires consideration of the agent–structure binary.
>
> (Chowdhury and Turner 2006: 303)

By describing the agent–structure relationship as a binary, Chowdhury and Turner seem to neglect the varying scales at which structure manifests itself. I suggest we think of structure as a nested hierarchy of forces acting at differing scales. Some structure is created by physical, geographically anchored infrastructure (like buildings and roads) and biophysical characteristics of the environments people use (like soil characteristics and energy resources). However, most structure is socially created by the institutions, groups, and individuals who exert power over the person or group or other entity of study.

Overcoming the limitations of discourses on the Himalayan environment

Recognizing the socially contingent quality of knowledge about human use of Himalayan environments should lead to more sophisticated discourses and improved interventions, but scholars have different ideas about how that might be accomplished. At the most basic level, scholars and practitioners will become more conscious of the assumptions they are making and this reflexive awareness should make them more careful and more precise. Roe (1991) argues that development planners will always need narratives to overcome the unavoidable uncertainty they face, so scholars should intervene to improve practice by presenting and supporting alternative or modified narratives which counter the ineffective, over-simple, or incorrect parts of the dominating narrative.

Leach and Mearns (1996b: 31–3) seek to improve Roe's limited goal of offering improved narratives by democratizing expertise, i.e., by combining research with policy-making and bringing all the stakeholders into the process of defining problems and designing solutions. The stakeholders they mention are scientists, policy-makers from government and donor agencies, local administrators, NGOs, local residents, and mass media. These actors are likely to challenge dominant narratives about human environmental relations in ways that reduce the mistakes that have been made in the past. Leach and Mearns answer three likely criticisms of their proposal: (1) conventional scientific interpretations will still be valued, but new perspectives will also be included, a process that will lead participants to view science more critically; (2) the differing amounts of power and unconscious biases of the stakeholders will make democratic interactions difficult to attain, but if those differences can be identified, confronted, and accommodated, the process will help uncover biases in ways that expand the awareness of all participants; and (3) the practical need for replacement narratives can be more effectively met by engaging all the stakeholders in the process than by relying on the construction of western scientists and policy-makers alone.

Forsyth's review of Himalayan crisis narratives (1998) argues that western scientists and development practitioners must integrate biophysical natural science with an informed social science, "to provide an awareness of social constructions of environment at the same time as a falsification of inaccurate statements unhelpful to environmental management or development" (ibid.: 108). Forsyth seeks to avoid the pitfalls of positivism and of post-modern relativism "by assuming that ecological approaches exist independently of human perception, but that all attempts to understand processes will reflect social or political agendas" (ibid.: 109). The way to do this, he claims, is to adopt critical realism, Cultural Theory, and "new ecologies" as theoretical approaches. Forsyth's main emphasis is to improve the quality of development by facilitating the collaboration between biophysical and social scientists, while also expanding their awareness of the socially contingent bases of the positions they support. He seems less concerned with providing development "subjects" with a voice in the development process, i.e., democratizing research and policy-making.

I see Forsyth's discussion of critical realism and the epistemological implications of socially constructed knowledge as his most important contribution. The characteristics of postmodernist thought are:

> a preoccupation with discourse and language, a suspicion of concepts of truth and falsity and of empirical testing, an incredulity towards "meta-narratives" like philosophical systems or Marxism, a distrust of the idea of progressive development of knowledge, . . . and an openness to difference and an advocacy of "local knowledge."
>
> (Sayer 1993: 320)

All these undermine claims to truth and can be easily interpreted to mean that all knowledge is relative and exists only within the heads of individuals. That the concern over the relationship between knowledge and an external reality has continually reappeared to roil the minds and debates of philosophers indicates to me that there is no definitive "answer" to this relationship. The recognition of the socially contingent nature of knowledge further exacerbates the ambiguity, so I believe we must *choose* our ontology and epistemology.

"Critical realism" is the philosophical position that there exists a reality external to our perceptions of it and that we can know this reality. Idealism, in contrast, maintains that we only know ideas; some versions, like Plato's, hold that only ideas are truly real, that a world of ideas exists outside our minds, and we only know these independently existing, immaterial ideas imperfectly. Other versions, like that of Berkeley, hold that we know only the ideas within our own minds and cannot know an external reality or even if one exists. Many permutations and interpolations between these positions have been presented.

Forsyth is not alone in decrying the ideological implications of discursive analysis. Sayer (1993), Bhaskar (1986), Lopez and Potter (2005) and Latour (2004) among others all argue for a realist epistemology and ontology. Sayer (1993) argues that postmodernists set up false dichotomies between absolutes, which he calls "pomo flips," refusing to recognize intermediate positions. The first "flip" is to limit choices between "foundationalism," the idea that there is an absolute foundation to knowledge which allows people to know truth, and absolute relativism, which makes talk of a truth or falsity absurd. Sayer notes the contradiction between this relativist position and the postmodern authors' behavior in arguing against realist or foundationalist positions: in a totally discourse determined world, foundationalism and relativism are equally valid.

In contrast, critical realists agree there is no absolute truth, but claim that we can have partial or relative truth and that an independent world intervenes to falsify our inaccurate ideas of it. Human theories and discourses can have varying degrees of practical adequacy because there is "some relationship between the structure of the discourse . . . and the structure of the material actions and objects which are its referents" (Sayer 1993: 326–7). If there were not this connection, we could change the world at will by imagining a new discourse.

Sayer's second false dichotomy is between "meta-narratives" or totalizing discourses like Marxism which construct a center (production and its social relations) which explains all social phenomena, and a world determined totally by a centerless web of heterogeneous, locally based discourses that cannot be applied anywhere but locally (ibid.: 332–3). Sayer argues that realists believe that patterns and structures exist which cause results across localities and are widely replicated, but that a plurality of such causal forces interact in ways that make "the activation of their causal powers unlikely to produce stable regularities or order . . . It does support the view that some structures (mechanisms, objects or whatever we care to call them) are more important than others in shaping *particular* outcomes" (ibid.: 333–4). Meta-narratives can only provide partial and contingent explanations, but they are indeed explanations. Lopez and Potter make the same point:

> We utilize empirical observation and human causal power in the process of experiment. But explanation does not directly follow from this. Rather inferential processes must be used to attempt to discern the underlying features of reality, to wit, the intransitive domain of real causal mechanisms. Both our observations and inferences may be faulty. Equally significant, . . . our observations and inference . . . are also limited. We view things from a particular perspective, which includes the limitations of our time and culture. Thus, knowledge evolves. Hopefully we produce truer and truer accounts of reality. But this "accumulation" of knowledge, this progress is not guaranteed in advance . . . Realist ontology certainly gives a strong place in science for empirical observation, but also insists that the truth of things-in-themselves does not usually lie upon the surface. It focuses on underlying structures.
>
> (Lopez and Potter 2005: 12)

Sayer's third false dichotomy is between privileging western culture and science as the sole base of "truth" and an absolute cultural relativism which defines "truth" as entirely a function of the discourse of which it is a part. If there are no standards by which differing interpretations of the world can be judged, then the powerful can dismiss the criticisms of the oppressed as merely a function of their discourse. Similarly, if discourses are totally relative, the discourse of racists must be considered as valid as those of human rights activists. Mawdsley's critical discussion of the Hindu right's attempts to use the struggle over the Tehri Dam in Garwhal to further their political goals (Chapter 8) exemplifies the belief that moral standards must guide one's interpretations. Her chapter, and indeed, all the chapters of this book implicitly, if not explicitly, assume a critical realist epistemology: the authors believe they are describing events and processes that occur independent of their perceptions, while at the same time seeking to identify the ways that narrative and discourse filter both their own experiences and those of the individuals and groups they study.

Cultural and epistemological context of this chapter

Since this book is an exploration of the relationship of culture to Himalayan environments, I want to make explicit the assumptions that I believe will facilitate the effectiveness of western-educated development practitioners. This chapter clearly privileges western science as its cultural base, but the science I seek to employ is one informed by critical realism. I see science as essentially similar to all human knowledge: the matching of a formal system of assumptions, concepts, and logic to our experience of the world. Science, however, is a more rigorous, self-conscious type of knowledge.

First, science is unique in its rigor. It considers only those aspects of human experience that can be unambiguously communicated, which usually means those that can be operationally defined. This means that we define "length" by the process we use to measure it. It also means that we exclude vast areas of rich, human experience that cannot be operationally defined. Second, scientists attempt to identify and make explicit their assumptions and definitions, they seek to be "reflexive." Much of this chapter, however, identifies how narratives and other explanatory models have remained unconsciously part of the thinking of the community of scientists and how these narratives have distorted that thinking. Third, science employs a very rigorous logic, ultimately mathematics, so the logic of the arguments scientists present is likely to be without flaw. The problems arise from the assumptions and the data or "facts" on which the arguments rest: this chapter gives abundant evidence that these data and facts are often virtually determined by the assumptions the scientist makes. Fourth, scientific experiments never prove a theory, but merely fail to falsify it. When a scientist investigates a theory, he or she will manipulate the mathematical expression of the theory to predict that performing experiment A will lead to observable phenomenon B. If the scientist does observe B, it does not prove the theory, because there may be other, better theories, which also predict phenomenon B and even do so more precisely. Einstein's Theory of Relativity is a clear example from physics. Indeed, based on recent history, the scientist should *expect* that current theory will be replaced at some future time.

If nothing else, this chapter shows how the assumptions and mental models individuals use to think about and interact with their environments determine their actions, and how the discourses of their culture reflect and reproduce these assumptions. The fallibility of these assumptions and models tends to push many into a "pomo flip" of either naïve realism or idealism.

I am arguing for the more mature and nuanced approach of critical realism. The incredible success of science-based technology in manipulating our environments makes the lay person and many scientists abandon the critical essence of science and naïvely assume their models of the world truly represent it. Development practitioners are even less critical of "scientific" explanations, as they seek the generalizations that justify their plans. Critical realism acknowledges the ambiguity and uncertainty in all knowledge and takes science's self awareness to a deeper level.

Despite the uncertainty, cultural bias, and relativity of our knowledge, I believe we need to *choose* a realist position: we can and do learn, albeit imperfectly, about a world that exists outside our heads. Moreover, our practical lives involve interacting with our environments in ways that work, and this confirms that we know reality with a considerable degree of accuracy. I find myself thinking of two anecdotes from the history of philosophy. The iconoclast Samuel Johnson claimed to refute Berkeley's idealism by kicking a rock; Kierkegaard compared Hegel, with his elaborate idealist philosophy, to a king who builds a magnificent palace, but lives in a shack by its side. We are all in our daily lives realists: we can and often do make errors, and we continue to hold unexamined assumptions, but the at least partial successes of our activities show that we do know "reality," albeit imperfectly.

The world is always more complex than our ideas of it, so we can never know it completely. This realization undermines the claim to realism, and this uncertainty is why I say we must *choose* to be realists. After all, human existence is itself ultimately ambiguous, and we continually choose how we give it meaning. Kuhn and Foucault identified *and gave name to* the ways our thinking is influenced by the sets of ideas that our social groups use to explain the world. By naming a discourse or a paradigm we become conscious of how our socially constructed meanings and assumptions affect our thinking and action. The reflexivity Forsyth urges is enhanced.

Conclusion

In this chapter I have sought to explain the reasons for the acceptance and rejection of the Himalayan environmental crisis narrative. The following is a list of conclusions:

1 Ideas about human impacts on the Himalayan environment have been based on narratives or discourses which tell simplified stories about an impending disaster and dictate the actions to mitigate and reverse it. The attempt to specify empirically the processes creating the disaster scenario and the failure of the programs derived from these narratives have led to a more precise and nuanced reassessment of the problem and to improved policies, like community forestry.
2 Human-induced environmental degradation, like the loss of biodiversity and scarcity of subsistence resources, is occurring, but the intensity of

these processes vary locally; in addition, these processes are being influenced by forces, like national forest policies and international aid, operating at temporal and geographical scales beyond the village.

3 Local villagers actively transform their environments to fulfill their subsistence needs by building terraces, planting trees, and creating social institutions to manage forests and other resources. The assumption of THED that their actions only degrade their environments is wrong.

4 The power of the crisis narrative comes from the benefits it brings to a multiplicity of institutional and individual actors operating at international, national, regional, local, and personal scales.

5 For example, in the post-colonial period, the US and her allies have guided Himalayan development by promoting fluctuating development discourses and by funding of their own bilateral agencies and the multilateral institutions to create and administer programs which conform to their development discourses. At the national scale, government agencies have adopted the crisis narrative to ensure the flow of international aid and/or to justify state control over resources.

6 At regional and local scales, "big men" struggle for dominance through electoral processes, by creating NGOs, and by adopting the current environmental discourse when it justifies their claim to development funding. Development planners and policy-makers at international and national scales use narratives to guide planning because they have little accurate information about the problems they are charged with solving and the narrative tells them what needs to be done. Their academic training gives them the expertise that justifies their claim to being a stakeholder in local land use decisions.

7 The policy-making process is a political rather than rational process, in which the finally implemented policy results from the balance between all the pressures the numerous "stakeholders" exert.

8 Scholars studying the roles that environmental crisis discourses play in academic studies and development practices argue that if scholars and development practitioners become more conscious of the distortions created by development narratives and other forms of discourse, they will be more effective.

9 In addition, acknowledging the social contingency of knowledge will make planners and policy-makers more willing to consider and give power to alternative interpretations of human–environmental interactions and other issues. What is key to such efforts is to structure the interactions so that all participants have the power to promote their interests, though creating and maintaining such a structure is a continual challenge.

10 The epistemological problems implicit in the claim that knowledge is socially constructed are challenged by critical realists who argue that we can know an external world, but only partially and imperfectly; when our

knowledge is grossly inaccurate, real events intrude to invalidate our interpretations and force us to revise them.

11 By taking a longer perspective, we can interpret the whole process of the emergence and rejection of the super-crisis narrative as an example of how idealized science works: a model is presented, its accuracy evaluated, and the original "theory" is modified or improved.

2 Healing landscapes

Sacred and rational nature in Nepal's Ayurvedic medicine

Mary Cameron

In memory of and deep appreciation to Siddhi Gopal Vaidya and Dr. Lokendra Man Singh.

Be glad and joyful in the Plants, both blossoming and bearing fruit,
Plants that will lead us to success like mares who conquer in the race.
Plants, by this name I speak to you, Mothers, to you the Goddesses.
He who hath store of Herbs at hand like Kings amid a crowd of men,—
Physician is that sage's name, fiend-slayer, chaser of disease.
Herbs rich in Soma, rich in steeds, in nourishments, in strengthening power,—
All these have I provided here, that this man may be whole again.
. . .
Reliever is your mother's name, and hence Restorers are ye called . . .
The Plants have driven from the frame whatever malady was there.

(*Rg-Veda* Hymn 97 "In Praise of Herbs," excerpts from stanzas 3–13; Griffith and Shastri 1973: 610–11)

Introduction

> What is often being argued . . . in the idea of nature is the idea of man; and this is not only generally, or in ultimate ways, but the idea of man in society, indeed the ideas of kinds of societies.
>
> (Williams 1980: 70–1)

In her book *Loving Nature*, Kay Milton (2002) asks a simple question: How do communities and individuals develop a protective and caring relationship with nature? Milton synthesizes ideas from religion, science, and psychology to advance, in one respect, Tim Ingold's theory that people's direct experience(s) with the natural environment acts alongside social constructs to produce deep emotional bonds with the environment that include its protection. It would not be surprising, then, to find that societies directly dependent on plants economically and medically[1] would have correspondingly high levels of plant-based knowledge and environmental stewardship, and that they may have evolved sacred and pragmatic techniques of conserving certain kinds of plants like medicinal ones. The Nepali people comprise such a society and Nepali farmers and others dependent on them commonly consult medical practices that rely on plants. Both Ayurvedic[2] and Amchi medicine extensively utilize plants to alleviate suffering due to illness and to rebalance the body's humors.[3] In this chapter, I detail how Ayurvedic doctors acknowledge and actualize a love for medicinal plants, *jaribuṭi*, and how their caring converges with 'scientific' language from biomedicine and environmentalism about nature. The use of plants in healing finds a receptive population among urban and rural Nepalis, and discourse about medicinal plants is generated in a variety of places.[4] In this chapter, I draw attention to the politics of conserving Ayurvedic medicinal plants within contexts of local engagement with nature and the environment that are common to lay Nepalis, specific engagements with medicinal plants by family-trained and academically trained doctors, and government and non-government organizations' efforts to protect medicinal plants.

The importance of cultural-based and development-driven conservation is, however, contradicted by political impulses that place medicinal plant expertise and conservation at the center of debate in the political economy of donor-driven health care in Nepal. First, the Nepali government has failed to regulate the flow of medicinal plants across its national borders and to resolve the ecological complications derived from illegal harvesting and trade. Second, concern about formal, institutional Ayurvedic education centers on the insufficient training students receive in medicinal plant identification and preparation, in contrast to informally and family-trained doctors who acquire extensive plant-based knowledge through their long apprenticeships with family members and other traditional teachers. Third, and contradicting the previous edition, recent efforts to make Ayurveda more professional and modern by regulating who practices and what plant-based medicines are produced ultimately limit practice to those with formal educational training or those who can prove the existence of three generations of healers in family lineage practices. Of concern to many, the legislation could erase medicinal plant knowledge (and inevitably health care delivery) in subsequent generations. I will address each of these issues in detail below. What is clear, though, is that in privileging biomedicine in its health care policies, the state neglects institutional development of Ayurveda. Autocratic licensing and medicinal plant conservation efforts are insufficient to counterbalance politics, capital interests, poverty, and biomedical dominance that work against medically-based nature conservation.

According to the World Health Organization, traditional medicine comprises the main source of health care for nearly 80 percent of the world's population in developing countries (Chaudhury and Rafei 2002). Of the traditional medical systems in South Asia, Ayurvedic medicine is integrated into the medical bureaucracies of most countries, and has continued for centuries to be a very popular system of healing throughout the region

(Dixit 1995; Himalayan Ayurveda Research Institute 1996). Ayurveda and its variants—for there is not one true or pure Ayurveda, but different forms and traditions—are considered to be the most common and popular forms of healthcare in Nepal, for Ayurveda provides, as many believe, a fundamentally sound theory of health and illness, one that is supported by people's extensive knowledge of medicinal plants and their uses. Furthermore, medicinal plants are part of social networks; lay Nepalis generously share their knowledge of plants with others—family, neighbors, and foreigners.

A central dilemma for Nepal is whether she can become modern while preserving and developing Ayurveda, which has not advanced like biomedicine has in the country. Conserving medicinal plants has become a kind of bridge between modern environmental movements and modern indigenous medical movements. Consequently, the language of developing Ayurvedic medicine increasingly turns on the status of medicinal and aromatic plants in the country, and on their efficacy as evaluated through allopathic medical standards, rather than on institutional and human resources development. In other words, developing plants and identifying their chemical properties within the language of western science has become more acceptable as a modern effort to preserve medical traditions than has developing its human resources. While medicinal plant protection may be a point that the Nepal actors can agree on, the ability of different groups to actualize that protection is limited, constrained as they are by either lack of resources and organization, or a lack of political will to adequately control illegal trade in a way that does not further impoverish the rural poor who are compelled to sell plants in less than environmentally conscientious ways. In this chapter, I thus engage with broader discussions on social and medical justice in an era of increasing biomedical globalization and diminishing natural resources.

Ayurvedic medicine and nature

> The Western doctors take into account the seed only, ignoring the soil. Hence they miserably fail in attaining the results desired.
>
> (*Journal of Ayurveda*, 1927; quoted in Sharma [1929] 1983: 144)

Ayurvedic medicine has historically drawn its therapeutic language from a set of natural images by which to situate humans within the phenomenal world, to further Ayurveda's 'natural' typologies of people, and to advance its theories of illness causation, progression, prognosis, and cure. Foundationally, classical Ayurvedic medicine detailed in the ancient medical texts (*Carakasaṃhita, Suśrutasaṃhita*, and *Aṣṭā ṅgahṛdayasaṃhita*) derives from Samkhya natural philosophy in theorizing that the living body is comprised of natural characteristics or *prakriti*, in the form of three *doṣas* or humors (*tridoṣa*), wind (*vāta*), bile (*pitta*), and phlegm (*kapha*), substances that flow throughout the body and exhibit "particular textures, colours, tastes, and locations in the body" (Wujastyk 2003: xlii). Though three in number, the *doṣa* are formed from five ubiquitous entities or elements (*pañcamahabhūta*) found in the phenomenal world: sky/space, fire, air, water, and earth (*akās, tej, bāyu, jal*, and *pṛthvi*). The three humors have identifiable qualities (*guṇa*) and tastes[5] that are in dynamic equilibrium with larger phenomena (seasons, planetary alignment) and with internal and external substances (food, metals, animals, plants) that also contain the same distinctive qualities and tastes (Zimmermann 1987; Durkin 1988; Zysk 1991). The living organism, constituted by the *tridoṣa*, interacts with an environment constituted by the five elements, six flavors, and six seasons. According to Ayurvedic theory, the two main causes of illness are faulty diet and faulty daily regimen, as these may disturb the "cooking" or "ripening" of *rasa*, the food-juice of life within the body (White 1996). Symptoms of illness are read by the physician as indicative of an excess, absence, or dislocation of one or more humors. Adjustments are made through dietary and other daily regimens and the consumption of plant-based medicines. The "ecology" and "economy" of the body and person in Ayurvedic thought also make the subject's daily practices, social relationships, and environmental surroundings integral to diagnosis and treatment (Nichter 1981, 2001; Zimmermann 1987).

Ayurvedic medicine has an articulated theoretical paradigm that is based on centuries of natural experimentation and observation, and has both formal and non-formal systems of training, a recorded *materia medica*, and a clinical tradition with a diverse range of therapeutic treatments. Health is defined in Ayurveda as a state of dynamically balanced *doṣas*[6] maintained by a congruence between the body and its physical and living environment;[7] this suggests a state of health best achieved, and indeed derived from, a "natural" and rural kind of environment, rather than an overpopulated and less "natural" urban one. If we look closely at linguistic and medical history, though, we find that employing an urban–rural or nature–culture dichotomy to classify illness and health may be overly simplistic. Ayurvedic practitioners of the past and the present recognize at least three distinct rural areas—the wild jungle or *ban*, the *jaṅgal*, and the *araṇya* (Zimmermann 1987)—delineated ecologies that nourish particular medicinal plants and contribute to the distinguishable types of human *prakriti*. According to Dr. Narendra Tiwari, a leading expert on Nepal's medicinal plants, *ban, jaṅgal*, and *araṇya* are often used interchangeably but in fact have different root meanings. *Ban* is the "wild forest" where tribes who cultivate the land live, and whose members are called *banbāsi*. Often the *ban* is species-specific, such as *sal ban* where *sal* or evergreen trees grow, or *śleṣmatak ban* where the Siva temple Pasupati is located in Kathmandu, though many of the trees are no longer there. The *jaṅgal* is a place where non-cultivating tribes live, like the Raute (see Fortier, Chapter 5 in this volume). *Araṇya* is a place where only non-human animals live, an ecology that is "too wild" for humans. This indigenous ecological typology is also applied to the classification of medicinal plants, including in the sacred texts.

We find in the discourse on Ayurvedic medicine in Nepal today that urban and rural locations are discursively constituted in ways that increasingly polarize the landscape of healing. Current and entirely urban-based efforts to professionalize Ayurvedic medicine in Nepal position the rural as a place where important healing plants grow, but proponents of modern professionalization also identify the rural as a natural place where non-formally educated healers, who are nonetheless believed to be lacking in their medical knowledge, practice. The absence of a formal education from an urban Nepali or Indian Ayurvedic college or university is thought to relegate these village practitioners to an inferior form of healing practice. From the perspective of a second group, those wishing to develop Ayurvedic medicine on the wave of global interest in indigenous knowledge (in particular, medical knowledge), as well as NGOs and INGOs seeking new ways to protect Himalayan biodiversity, the rural is positioned very differently. For this group, the rural is an idealized original source of medicinal plants, a place of dietary-based humoral balance and sacred conservation practices that stands in superior contrast to materialistic, crowded, polluted and imbalanced urban places. Similarly, the current interest in Ayurvedic medicine among urban educated elites comes from Ayurveda's association with "natural" efficacy rather than from its ancient heritage. Thus, Ayurveda is an important source for how the Nepali people understand and interact with nature.

Ayurvedic doctors give different reasons for choosing their profession, but many do express an early love for plant life and feel that one reason Ayurvedic medicine is popular is because people are familiar with plants and are comfortable with their use in health care. Lay people and physicians alike find that Ayurveda is particularly suited to Nepal because, aside from being an efficacious system of therapies, the natural philosophy behind it provides a persuasive model for the physical and living world, and has shaped people's relationship to nature in many ways: as a way to think about farming, as a way to think about making the body strong and healthy, and as a way to realize an enduring respect for the god-given world of powerful, healing plants. Ayurvedic medicine is thus seen as possessing cultural common sense. This combination of phenomenology, metaphysics, and pragmatism shapes people's sense of moral duty to nature and the environment.

History of Ayurveda in Nepal

Up through the contemporary period, Ayurvedic medicine has been one of the most common forms of health care in South Asia. In Nepal, the importance of Ayurvedic medicine before the middle of the last century is evident in several historical facts. The Rana courts in Kathmandu maintained a hierarchy of Ayurvedic family doctors and established important Ayurvedic institutions during their reign. Each Gurkha military unit took an assigned *baidya* to England in the first decades of their British employment. And from oral histories we learn that a large majority of elderly Nepali citizens know herbal medicinal preparations and can identify numerous, common medicinal plants (Cameron 1986; Himalayan Ayurveda Research Institute 1996). In village and urban life, healing was achieved through people's opportunistic resort to self-cure, and through visits to *baidya*, herbalists, shamans, Tibetan doctors, wandering ascetics, and priests—a pluralistic healing landscape typical today that now also includes biomedicine (Adams, 1998; Burghart 1984a; Parker 1988).

In Nepal, formal Ayurvedic education began with King Chandra Shamshere Rana, who in 1928 established an educational fund for students to study Ayurvedic medicine in India. They completed their studies in five years and returned to Nepal to inaugurate the teaching hospital at Naradevi. Its mission was to advance the teaching of Ayurvedic medicine and to serve patients free of charge.

When Nepal opened its borders to the world in the 1950s following the removal of the Rana rulers and the re-ascension of the Shah lineage, the contemporary era of development (*bikās*) began. The Shah rulers strongly supported modern scientific medicine. According to former royal Ayurvedic doctors with whom I spoke, King Birendra gradually removed Ayurveda from the palace by replacing the Ayurvedic practitioners with allopathic doctors and introducing a retirement age of 60, which encompassed virtually all of the Ayurvedic doctors affiliated with the palace. Pressure to modernize the country was met with the introduction of European health care development in the form of immunization programs, public health, the establishment of a world-class biomedical college and auxiliary schools, and a rapidly expanding pharmaceutical industry (Justice 1986).

The formal educational system was restructured in the second half of the twentieth century by the Shah kings, who vigorously championed modernist ideology in slogans such as "education to meet the needs of the people." Finding Ayurveda non-modern, reformers restructured its medical curriculum, too. The Sanskrit-based curriculum was eliminated, and the four-year degree was reduced to a three-year program in line with other IA degree programs. The Ayurvedic degree was renamed "Proficiency in Complementary and Traditional Medicine," a title which altogether dropped the word "Ayurveda." Supervision of the main institution of formal Ayurvedic education, Naradevi Teaching Hospital, was moved from the Ministry of Health to the Institute of Medicine in the Ministry of Education in 1972, when modern allopathic medicine was introduced as a discipline and a degree program. Today, the curriculum for the BAMS, Bachelors in Ayurvedic Medicine and Surgery, is five and one-half years; it includes one year of clinical rotation, and the amount of biomedically-related subjects taught comprises about 30 percent of the entire curriculum. However, the superintendent of Naradevi and many others cite an extreme lack of academic manpower as a major barrier to the professional development of Ayurvedic medicine in Nepal.

While these changes in formal professional institutions were taking place in the second half of the twentieth century and the beginning of the twenty-first century, what was happening to the largest group of Ayurvedic practitioners, the traditionally and family-trained healers? There are no national figures on their number in Nepal. Clearly, though, within a society of native botanists, these women and men *baidyas* possess some of the most intricate understandings of plant life. A recent survey of six Village Development Councils in Gorkha district identified 455 traditional practitioners who were providing people with most of their primary health care outside the family, and who were using locally available parts of approximately 250 plant species belonging to 86 families (150 of which are mentioned in Ayurvedic classics; Himalayan Ayurveda Research Institute 1996). Though this is an isolated study, ethnographic evidence from other areas indicates a culture heavily dependent on village healers using medicinal plants.

The fundamental Ayurvedic principle of balancing the body's dynamic humoral essences with diet, proper behavior, and medicinal plants, has noteworthy consistency across contemporary populations (Dixit 1995; Himalayan Ayurveda Research Institute 1996). Beginning in 1802, botanical, ethnobotanical and medicinal plant classification in Nepal's diverse ecosystems and ethnic communities confirms widespread and semi-uniform use of plants for health care purposes. In the contemporary period, this has been noted extensively in numerous studies and reports on Nepal's environment and its medical systems, though scholars have not always made the link to Ayurveda. Very often during my own village-based research, friends would pick the leaves and pull the roots of wild plants during visits and walks, to be administered later in home and community medical treatment.

Today, the health and medical systems in Nepal can be grouped into four categories: the home-based system; the faith healing system; so-called "traditional medicine" that includes Ayurvedic medicine, homeopathy, Tibetan medicine, and Unani; and modern allopathic medicine, with its vast network of hospitals, clinics, medical and nursing schools, pharmacies, and practitioners.[8]

Nepal is also home to a state institution that harvests medicinal and aromatic plants for use inside the country and for export. Established in 1981 under the Ministry of Forestry, the Herbs Production and Processing Company's (HPPCL) exports are comprised of approximately 20 percent medicinal plants, most of which go to European countries.[9] Many plants are grown on its farms in the Tarai, and the staff negotiates with private companies and individuals to collect and deliver plants to regional offices. HPPCL supplies some raw materials to the Vaidyakhana, including *śilājit*, a mineral pitch that is collected in small quantities in the Himalayas and is the main ingredient in a popular food supplement. The government of Nepal also grows medicinal plants under experimental conditions.

Private Ayurvedic drug production companies in Nepal include Gorkha Ayurveda and Dabur Nepal, as well as hundreds of small family Ayurvedic businesses that produce primarily for their communities, some of which may escape the purview of the Department of Drug Administration that oversees drug distribution.

Ayurveda in Nepali culture and practitioners today

Baidya in rural villages diagnose physical and mental maladies, and based on those diagnoses, they procure, prepare, and administer plant medicines. *Baidya* develop their skills from their own families, nearby kin, and from the larger community. They are part of what makes Ayurvedic medicine more acceptable and accessible than modern biomedicine to the Nepali people, practicing in ways embodying Ayurveda's holistic representation of the human body and its non-medicalization of the human life cycle. Other reasons people use Ayurveda include its non-invasive diagnostic techniques, the value it places on patients' social environment in illness treatment, its use of symbols of balance, mutuality, and cooperation, and its free service to the poor. The broad integration of Ayurvedic ideas into Nepali culture extends to lay people who apply humoral principles in self-cure, too.

In one of Nepal's best-loved epic plays, *Muna Madan*, written in 1936 by Laxmi Prasad Devkota, the main character, Madan, falls ill while returning to Nepal from Lhasa, Tibet. His traveling companions abandon him, advising him to repeat the name of the Lord, as he was surely going to die. After they leave, Madan is helped by a Tibetan man who takes him to his home, where

> [he] laid him down in wool. There he gave him water and tended him kindly; he went out to search for a forest herb, crushed it and gave it to Madan. Then he gave him milk from his yak cow and made him strong again.
>
> (Hutt 2001: 38)

Madan returns home to find that his beloved wife Muna, who had pleaded with him not to travel to Tibet, has died from loneliness and his mother passes away shortly after seeing her only son for the last time. Madan's health declines and he exhorts a doctor to consult his medical books:

> Read Charaka, turn the pages of Sushruta, where is it, tell me, this anguish of heart? The disease that afflicts me is life – cure that! The doctor looked and understood; he never came again. Wherever the heart's anguish might be, medicine cannot find it.
>
> (ibid.: 54)

The emotional turmoil that befalls the man who seeks fame over love cannot be cured even by the power of Ayurvedic medicine. Both Devkota and the play are familiar to millions of Nepalis, and while Devkota broke with long-standing literary traditions by using language and imagery familiar to common Nepalis, he also tapped into a cultural identity replete with medicinal plants and learned Ayurvedic healers that would appeal to his readers.

Practitioners of Ayurvedic medicine in Nepal come from different economic and geographic backgrounds, though most of them are from the upper castes. Colleagues I have worked with include non-literate rural *baidya* who have extensive knowledge of local medicinal plants and live on the margins of poverty like many of their rural neighbors, and urban physicians with advanced Ayurvedic degrees from India who are successful proprietors of clinics for foreigners during the day and healers at community Nepali clinics during the early evening hours. There are those who work primarily in administration in the Ministry of Health, and those who see patients throughout the day at family-based clinics that may be hundreds of years old.

Traditionally trained doctors belong to one of three groups. The first and most well known are those trained by family members with practices going back many generations. These families tend to be high-caste Newari and urban-based, and many have had family members appointed to Nepal's royal families as doctors. They often have extensive genealogies of their Ayurvedic practice, with manuscripts on diagnosis, treatment, and medicinal plant pharmacology dating back hundreds of years. They tend not to divulge family healing secrets and rarely have they trained non-family members in the art and science of healing. Some of them prescribe medicines only they have prepared, refraining from using mass-produced and Indian medicines. Women from these families may also serve as compounders and dispensers of medicine and as nurses, though rarely do they diagnose patients (for a fuller description of women in Ayurveda, see Cameron 2010). The professional stature of this exclusive group of *baidya* is based both on their success as doctors and the reputation of their lineage ancestors.

The second group of non-formally educated *baidyas* consists of people with apprentice roots that are contemporary, having been trained in families without long histories of Ayurvedic healing, or by non-family members such as seers and renowned religious figures. These *baidyas* practice in rural and urban communities and their professional stature is based mainly on their success at healing, though some claim the eminence of their teachers as evidence of their own abilities. This group meets the health care needs of most Nepali people, and is more diversified by caste, ethnicity, and gender than the first.

In the name of modern science: controlling Ayurveda rationally

> We must preserve the knowledge of the traditional practitioners, but we must also protect the people from fraud.
>
> (Dr. Ram Chanda Pandey, Registrar of the Ayurvedic Council, 8 March 2005, discussing the 2045/55 Ayurvedic Council Act)

Given its place in Nepal's medical history, it is not surprising that Ayurvedic medicine—its practice, its practitioners, and its patients—would become enmeshed in the country's health politics. In the case of Nepal, due to its relatively small size and location between India and China, the country has been a well-researched recipient of international development aid. In the field of health care "modernization" one is left with the clear impression that biomedicine is a strong—and growing—force in the country, powerful enough to determine the standards by which Ayurveda is measured. The presence of modern scientific ideology alongside an equally formidable theoretical system like Ayurveda raises a set of questions about the nature of that relationship. One manifestation of science power coming from a small but influential sector of the Nepali state would seek to control Ayurvedic medicine by means of a standard that bears little relationship to how individuals become healers, a standard that recognizes only one form of higher medical education and that gives little rational weight to the non-formal and traditional educational processes that have maintained Ayurvedic healers over the generations.

Pressure to regulate Ayurvedic medicine is an outcome of the larger goal to modernize health care in Nepal, and produces rhetorical constructs of Ayurveda as a non-empirically based traditional medical system worthy of support only as an alternative to modern medicine. Many health officials, themselves from modern scientific or medical backgrounds, do not consider the village practitioners to be true healers, and some have even publicly described formally educated practitioners as backward. One Health Secretary ended a discussion with me about the Ministry's plans for advancing Ayurvedic medicine with skepticism about the village healers, suggesting that they should not be called *baidya*, in spite of the fact that they practice in ways similar to professionally trained doctors. Still, government officials who characterize Ayurvedic medicine negatively may lose support from a population that cherishes its medical tradition. Dr. Ram Yadav had twice been appointed Health Minister, but during his first term, according to Dr. Lokendra Man Singh (Nepal's leading Ayurvedic physician until his death three years ago), he was quoted as saying that Ayurveda is like "the cart being pulled by the oxen," whereas allopathic medicine is like a "supersonic jet." With those words, he lost his bid for reappointment, though he was later reinstated. Politicians are requesting more Ayurvedic facilities for their districts and for the nation as a whole, as they recognize that Ayurveda fills a critically important primary health care need and they see how India benefits from Ayurveda's increasing global popularity (see Langford 2002, on modernizing Indian Ayurveda).

Efforts by the Nepali government to regulate both allopathic and Ayurvedic medical practices implicitly involve broader and profound concerns over the status of modern science practices, including the role of science in society, the definition and ownership of science, and the basis of scientific knowledge. Strategic claims of being the true followers of scientific healing are found on both sides, with the Ayurvedic community claiming its millennia-old systematic accumulation of knowledge rivals the knowledge produced by the relative newcomer, post-Enlightenment biomedical science. As scientists, Ayurvedic doctors believe it is possible to make steady progress in the sphere of human knowledge; that methods and goals are, or should be, ultimately identical throughout the sphere; and, that it is possible to derive the structure of the laws of nature from a single set of clear, abstract principles and concepts, if they are applied correctly (Berlin 1981). Like modern scientists, practitioners consider Ayurvedic principles to be universal, and they apply a method of diagnosis and treatment consistent with those principles.

Ayurvedic physicians point out that contemporary experimental trials of Ayurvedic medicinal preparations are inherently flawed because they do not test the actual mixtures prescribed in the ancient texts, and instead focus their search on a single active ingredient. This approach contradicts the important Ayurveda principle of combining many plants and their many parts, as well as the principle of *yogbai*, which is the addition of a plant that accelerates and enhances the main drug's action. Furthermore, the ecology of a plant is understood to affect its constituent parts and its chemical configuration, and therefore it is important to properly select plant source locations for experimentation. Finally, to measure efficacy is to measure a state that Ayurvedic doctors consider subjective and not measurable by the tools of modern scientific and medical experimentation.

On the other hand, some Ayurvedic doctors explicitly embrace the language of contemporary science when discussing Ayurveda, as either a legitimizing discourse or as another kind of science parallel to non-western Ayurvedic science. An example of the former is found in discourse emphasizing the importance of knowing the chemical basis of the drugs and assessing their efficacy. Here, modern science is seen to work in the service of Ayurvedic medicine, as a tool by which Ayurveda can supplement its own rich knowledge of how medicines work. Still others strategically exploit the language of science in an effort to legitimize the profession. "Scientific and Reliable Ayurveda Service is the Foundation of Healthy Life" is the motto of the Dhanwantari Ayurveda Hospital and Research Center (DHAREC), newly opened in Kathmandu by a group of Naradevi graduates, and recently put under the supervision of Dr. Sarita Shrestha, a Nepali graduate of Banaras Hindu University. It claims to offer both Ayurveda and modern diagnostic tools, with treatment provided by Ayurveda specialists. One of its objectives is "to perform scientific research in the field of Ayurveda and its impacts to the world health [*sic*]" and to offer training programs so as to provide "scientific knowledge of Ayurveda to traditional Ayurveda healers." The juxtaposition of academically trained doctors to non-formally trained healers is an attempt to bring together two communities of practitioners that have been recently divided over the issue of professional registration, though it does so by uncritically accepting the rationale by which they were separated—that knowledge gained through formal academic training is the standard by which all other forms of knowledge should be evaluated.

Some formally trained Ayurvedic doctors trying to establish new urban practices tend to discredit informally trained doctors that keep healing secrets inside the family. The traditional practitioners, well aware of the criticism, counter with their own concern that knowledge in the wrong hands would be harmful to patients.

Despite the absence of national data on Ayurvedic practitioners, the Nepali government has recently sought to regulate the practice of Ayurvedic medicine in the country. Succumbing to criticisms of Ayurveda as non-scientific, and bolstered by reactionary fears from the allopathic community and like-minded development officials that plant-based medicines may be toxic, the Ministry of Health, through the Department of Ayurveda and the Ayurvedic Council, sponsored legislation in 1988 that sought to regulate the practice of Ayurvedic medicine by mandating the registration of all practitioners. The Ayurvedic Council Act, eventually passed ten years later in 1998 and implemented in 2001 through the Ministry of Health, may have far-reaching implications for the practice of Ayurvedic medicine in Nepal, for it seeks to license practitioners on a very narrow set of criteria (Cameron 2009a).

According to the Act, individuals who want to practice Ayurvedic medicine must register with the Department of Ayurveda by proving their training with a degree or certificate in Ayurveda from a recognized institution, or, if traditionally trained and over 50 years of age (no one younger can apply under this provision), by documenting their status with a letter from their Chief District Officer stating that Ayurvedic medicine has been practiced in the applicant's family for a minimum of three generations. Equally autocratic are the restrictions placed on the production of medicinal plants, which, if read literally, would prohibit every Nepali citizen except those formally trained in Ayurvedic medicine from producing plant-based medicines.[10] The traditional *baidyas*, most of whom do not have formal institutional training in Ayurvedic medicine, now find themselves in open dispute with their professionally trained colleagues, for they see the new regulation as threatening their very livelihood. Nonetheless, recognizing the potential loss of centuries of botanical medical knowledge should the regulations be widely enforced, the traditional community has responded with a certain degree of openness about both their formulas and their preparation methods. Sharing medical knowledge during open monthly symposia is new to the traditional community.

Intending to protect the people from dangerous medical practices, the state is declaring one kind of knowledge superior to all others—the knowledge obtained in institutions of higher education. Consequentially, the mechanisms for regulating medicine have shifted from local community control to the state through its oversight of educational, health care, and modern development institutions.

Physicians and plants in Ayurvedic education and health care development

People involved in Ayurvedic medicine agree that medicinal plants are central elements in two issues at stake in Ayurveda's future in Nepal, namely the development of Ayurvedic institutions and human resources, and the environmental protection of medicinal plants that are the basis for its therapies. For both formally and non-formally trained physicians, caring for and preserving plant life is equivalent to caring for and preserving human life, although non-formal routes of knowledge transmission more thoroughly teach about plants than do the formal institutions.

For physicians, the human body is more than a part of nature, it is a microcosm of the physical and living natural world, undoubtedly of great complexity and distinct in its possession of consciousness and spirit, but nature nonetheless. The physical body sustains a balanced equilibrium of humors through conscious and intelligent effort on the individual's part to recognize and utilize the material, natural world around her. Furthermore, one is predisposed through one's *prakriti*, one's own characteristic nature, to respond to the natural world in distinctive ways. Just as the medicinal quality of a plant will vary according to the soil in which it grows, so too the body and the individual will grow in organic connection to their nature and the environment they live in. Thus, the doctor must examine and evaluate what is called the "field," or the full, integrated body of the patient, view the present disorder within that multi-dimensional site, and decide treatment accordingly. Like the biomedical doctor who crafts a treatment from synthesizing symptom and therapy (pharmacological, etc.), the *baidya*, too, must connect pharmacological knowledge to symptom and field. This requires medicinal plant knowledge, typically gained in rural apprenticeships, urban lineages, or formal Ayurvedic pharmacology courses called Dravyaguna. However, the increasing integration of allopathic medical ideas into Ayurvedic medical education has meant that students spend less time spent in courses on medicinal plant identification and increasingly more time in modern biomedical and related subjects. Part of the problem stems from the fact that as Ayurvedic education becomes more urban and more integrated with allopathic medicine, the opportunities to develop plant identification skills become limited.

Dr. R. R. Koirala, a prominent member of the Ayurvedic community and a former teacher at Naradevi College, is critical of the current lack of attention to field experience in Naradevi's Ayurvedic medical curriculum:

> It's not a matter of making comparisons between now and the past. The vision at the start of the institution was defective. When a student graduates he does not even know how to grind a plant. When the students collect the plants, do they collect the right ones or not? They read about beautiful plants in the books, but if they spend about a month

> in the field and observe how the plants grow they'll have even better knowledge.

Dr. Koirala himself attributes his knowledge of plants not to his education at the highly esteemed Banaras Hindu University, but to his life with a particularly gifted aunt as a youngster in village Nepal:

> As a young farmer I had a lot of experience in these things. My elder aunt showed me so many things . . . Although I do not remember everything, I do remember that I was always with her when I was ill. When my brother was severely ill for more than fifteen days, I remember collecting some plants with her. We collected leaves of specific colors and at specific times of the day. People like her had much more knowledge than our formal knowledge in Ayurveda . . . Today's students should experience this.

For this well-respected teacher and doctor, direct experience with plants in their natural environment by doctors-in-training is very important to their development as doctors.

Some physicians are filling gaps in their own education by spending time in apprentice relationships with knowledgeable doctors. Dr. Sabitri Thapa is concerned that although her Bangalore education was excellent in pure Ayurveda, she learned only the local names for plants—not the Latin and certainly not the Nepali names. So she finds herself unable to teach effectively about medicinal plants until she completes the informal tutoring she is receiving from her colleague, Dr. Narendra Tiwari, a leading Ayurvedic botanist.

One innovative approach taken by two highly successful physician-teachers interested in promoting and helping develop Ayurveda in Nepal involved incorporating Ayurvedic pharmacology into allopathic medical education. The main goal of Dr. Lokendra Man Singh and Dr. Tiwari, both members of the Naradevi teaching staff, was to increase biomedical pharmacists' use of medicinal plants. During one meeting I attended between the principals at the Institute of Medicine campus in Maharajganj in April 2000, the doctors introduced to the pharmacy faculty the idea of incorporating Ayurveda in the BPharm degree, including identification of major medicinal plants, their uses in Nepal, and Ayurveda's crude drugs standards and manufacturing methods. They encouraged the inclusion of the principles of Dravyaguna (Ayurvedic pharmacology) in the otherwise allopathic curriculum, to constitute a minimal 5 percent of the entire pharmacology graduate program. Pharmacy graduates would then be employable in Ayurveda dispensaries, as well as in allopathic pharmacies.

Another kind of rational management introduced into Ayurvedic medical organization is found in various WHO projects. One of the largest and most influential organizations to promote traditional medicine globally, WHO advises on how best to convert non-western science and non-scientific practices into western scientific ones. Through their partnership with the government of Nepal and under the direction of the Department of Ayurveda, WHO supports projects under their general guidelines of quality assurance in medicinal practice and drug manufacture; one such project I partially observed was the development of an essential drug list, described below. A second involved research on hepatitis A to evaluate the potential toxicity of certain traditional applications. Underlying the seemingly useful production of drug lists for countries like Nepal rests a particular kind of ordering of medicine intended to alleviate fear (of toxicity or fraud, both only speculated about and rarely documented) by making Ayurveda's use more regulated (and hence more rational). WHO has also provided funding to stock rural pharmacies and to enhance the Vaidyakhana's production system.[11]

In the last week of July 1998, during the early stages of my research on Ayurveda in Nepal, I attended a two-day workshop at the Plaza Hotel in Kathmandu entitled "Workshop on the Preparation of Essential Drugs List from Medicinal Plants." It was attended by 14 practitioners and organized and sponsored by the Ministry of Health, the Department of Ayurveda, and the WHO. The participants first established several principles for the medicinal plant list. They agreed that they would not list plants that are rare or endangered, and would include on the list plants that were easily identifiable. The list would also contain some mineral salts. Finally, the list should be local, not "foreign"—meaning that it would not include plants from India. As they proceeded, they based their inclusion criteria on the cumulative experience of clinicians and the recommendations of the classical medical texts, and explicitly not on laboratory identification of a plant's active chemical make-up. Modern scientific techniques were briefly discussed in the context of "animal experimentation" conducted at Banaras Hindu University. Dr. Singh voiced strong dissent to including drugs verified in this manner, claiming there was no need to know the chemical constituents of drugs. As he put it, "We all drank our mother's milk without knowing the chemical make-up and it was good for us!"

Science has come, symbolically, to represent in Nepal the modern present, in contrast to those knowledge systems like Ayurveda that originated in the country's pre-modern and traditional past. In justifying the virtual exclusion of Ayurveda from national health care plans that are tied to development funding, for example, the refrain is the same—Ayurveda is not scientific, so it cannot be included in a substantial way. The arguments that have been made over the years for claiming that Ayurveda is, in fact, one of the earliest and most enduring forms of medical science, in the broader sense of that term, are rejected by the allopathic community in Nepal. They note that Ayurveda fails to test its hypotheses and does not, in their erroneous view, allow for revision based on experimentation. The natural and historical experiment that is Ayurveda today seems to be unavailable to the allopathic community's vision of legitimate medical knowledge.

Ayurvedic development is plant conservation

A point of national pride for Nepalis is their country's unique geography, ranging from the flat plains of the southern Tarai to the high northern Himalaya. Ayurvedic doctors consider the unique ecology to be the foundation for the rich resource of medicinal plant variety. But they are increasingly alarmed at the state's inability to protect medicinal plants or to successfully develop the resource commercially for national and international trade. At the same time, India is believed to exploit Nepal's medicinal plant resources, often selling processed and packaged Ayurvedic medicines back to its own citizens.

Like many of her developing country peers, Nepal's environmental conservationism is an effort jointly approached by a patchwork of government agencies, NGOs, and INGOs, including the Ministry of Forestry and Soils Conservation, the World Wildlife Fund, ICIMOD, International Development Research Center, Canadian Centre for International Studies and Cooperation (CECI), smaller agencies funded through international donations, and academic departments at Tribhuvan University. Interestingly Ayurvedic physicians sometimes serve as botanical consultants to these conservation organizations. Organizations tend to share a technologically and scientifically oriented approach to environmental conservation that draws from international practices and standards, with many institutional agents having been trained in western countries. Indeed, the burgeoning interest in medicinal and aromatic plants (MAPS) conservation was made delightfully evident to me during a trip to visit the head physician at the zonal Ayurvedic Hospital in Dang, Dr. Devi Bhandari. I had been delayed by about six hours at the airport in Kathmandu, and upon arriving in Dang, I was kindly offered a ride to my final destination by some fellow travelers. The driver introduced himself as the regional judge, and I introduced myself as an American anthropologist studying Ayurvedic medicine and culture. A backseat passenger was interested that I was studying Ayurvedic medicine. He enthusiastically noted that "as we drive along there are medicinal plants on both sides of the road. Everybody uses Ayurvedic medicine in some form, in their homes. People recognize the plants." He sounded like a botanist, and when I inquired he explained that he was a forester studying for a doctorate in England. In fact, two of the three Nepali men in the backseat were doctoral students from England. One was studying the impact of forestry management on marginal landholders and the other was recording local people's knowledge of medicinal plants. We exchanged cards and I got out in front of the large compound that is the regional Ayurvedic hospital. As I watched them drive away, I smiled to think that the carload of Nepali plant lovers were equipped to have had a knowledgeable conversation with Dr. Bhandari about medicinal plants. And I also wondered if they were able to retain their local understandings of human–plant relations while being indoctrinated into the science of western environmentalism and western botany. I remember deciding they likely were comfortable with the fit. I am not certain I would agree with that conclusion today, ten years later.

Medicinal plants conservation in Nepal is unique because of the Ayurvedic medical community's involvement in it. This makes sense due to their professional dependence on the quality and quantity of medicinal plants in the country. Data about non-timber forest products exportation, for example, which are mainly comprised of MAPS and range from official estimates of $4.7 million in royalty to nearly $27 million unofficially, interest the medical community as they monitor in particular annual exports to India (alone estimated to be $18 million; Kanel 2000).

At a symposium on Ayurvedic medicine I organized in August 2000 at the American Center in Kathmandu, a group of Ayurvedic doctors addressed the topic of plant exportation from a number of different directions. Dr. Tiwari expressed the community's concern about the neglect of Nepal's plant resources.

> We are very rich in biodiversity, especially the diversity in plants. We are ranked 27th in the world in biodiversity. But we have to see it as a potential for providing revenue for us, as a major economic resource of the country. We are said to be rich in water resources but we have to buy a bottle of mineral water for 20 rupees. Similarly, we are rich in *jaribuṭi* but the government has no interest in developing their production. With difficulty, a Vaidyakhana at Singa Darbar has been established. The local manufacturing units of Ayurvedic drugs in Nepal fulfill only 15 percent of the internal demand while the remaining 85 percent is imported from India. The maximum portion of our medicine is from the Dabur Company. Dabur Nepal also produces the maximum portion of the finished products that are exported. This company does not belong to the government but to the private sector.

No one knows precisely how many plants have medicinal value. Contrary to the oft-cited number of 700 medicinal plants in Nepal, Dr. Tiwari notes that he himself has identified 1463 that are used by people, and he believes that his data may be incomplete. In his description below of the cycle of debt through which poor villagers become involved in the legal and illegal trade of medicinal plants, he strongly advocates that Nepal become more self-sufficient in the area of medicinal plant conservation, warning that the current state of affairs leaves Nepal vulnerable to global forces with potentially devastating outcomes:

> The government states that we should encourage the growth of the medicinal herbs. But to which species should we give priority? In Nepal, we haven't been able to utilize the 1463 species of medicinal plants. In China there are some 30,000 species but there are only 5000 species that are

traded. Here there aren't even 700 species traded. There is no research and development program on medicinal herbs that could investigate which of the 1463 species are vital or potentially beneficial for us. There are no organized government bodies to promote their use. The tendency we have is that if an American comes here and discovers that we have *taxus*, then we start to cut the plants and supply them to America. And what economic impact will it have when America starts using their technology to synthesize *taxus*? Take another case of *raulfia* or *sarpaganda*. There was a time when its roots were excessively supplied to the foreign countries. Now the government has banned its collection. When we have abundant natural resources, we seem to sell them away hoping to become a millionaire overnight. We never think about the long-term impact . . . What we need to remember is that most of the species of these herbs are going out through wild collection. In the case of wild collection, poor villagers who have difficulty for their everyday lives collect the plants. First, they take loans from moneylenders who then demand a certain quantity of plants within a given period of time. Hence they create their own rates, not the market. What happens in the villages is that they usually have to collect the plants from the common lands. So everyone wants to collect the plants first no matter if they are immature. They do not have time to wait till the plants mature, bear fruit and drop seeds that will grow into new plants. What will be the quality of drugs made from these immature plants? If one genus or species is lost from the wild, it is like losing one of our precious jewels.

The National Apothecary Dr. Tiwari mentioned is Singa Durbar Vaidyakhana, established as Nepal's first organized medical structure nearly 315 years ago during King Pratap Malla's reign. Originally established in the royal palace, it was moved to Hanuman Doka by Prithvi Narayan Shah, and then eventually moved to its present location by Prime Minister Juta Shumshere Rana over 80 years ago. Its primary function is to produce Ayurvedic medicines from Nepal's medicinal plant resources using classical methods, and to obtain patents for various formulas. There is an Ayurvedic doctor available for consultations at the facility. On 23 July 1998, after a week of preparations, an official ceremony was held to mark the establishment of the Apothecary as a semi-autonomous unit, separate from the government and to be run like a private enterprise. One of those who spoke at the ceremonial opening of the Apothecary was Dr. Krishna Kant Adhikari. At that time, he was the President of the Board for Singh Durbar Vaidyakhana. He told me in a series of interviews later that he became interested in Ayurvedic medicine at ten years of age, when his mother became seriously ill from postpartum complications. The family brought her to Kathmandu, a six-day walk, but she eventually died after one year. After that experience, he decided that he "must know why people suffered from illness." He apprenticed with a neighbor for a while, but then discovered there were no good Ayurvedic medical programs in Nepal. He studied Sanskrit at Banaras Hindu University when he was 16 years old, and then attended the Ayurvedic college there. He returned to Nepal at age 23 and began government service in villages that continued for 18 years. He returned to Kathmandu and taught at Naradevi for 10 years then founded the Department of Ayurveda in 1982 with the support of King Birendra. As its Director-General, he received a WHO grant for developing Ayurvedic medicine. Dr. Adhikari worked his entire life practicing and advocating for Ayurvedic medicine, and in his retirement others continue to press for personnel and other resource support from the government.

However, Ayurvedic medical development currently does not center on institutional and personnel development, but rather on issues related to plants and to plant-based medicines. In emphasizing the central importance of plant conservation in developing Ayurveda in Nepal, though, we must acknowledge that this reflects the understanding Nepalis have of medicinal plants and the value they place on them, on the one hand, and the state's inability to act autonomously of international and global pressures to modernize its health care system, evident in the state's lack of significant attention to Ayurvedic educational and institutional resources.

Meanwhile, in the homes and clinics of numerous Ayurvedic practitioners, conservation of plants continues. Many Ayurvedic doctors today hold workshops and manage plant collection and drying rooms in their clinics and homes to educate local communities about plant identification and conservation. Nathi Maya Dhoubadhel, a traditional practitioner from Bhaktapur, told me of her concern that the plants could no longer grow in the urban environment because of burgeoning housing development and the use of chemical fertilizers. At her home, she grows many plants in beautiful hand-made pottery from Bhaktapur.

An important way that the Ayurvedic community partners with local communities in medicinal plant conservation is through programs designed to identify local plants, utilizing people's knowledge of plant-based medicine. The staff at the Ayurvedic Health Home (owned and operated by Dr. Koirala), and the Dhanwantari Ayurvedic Hospital and the Devima Rural Ayurvedic Hospital (supervised by Dr. Shrestha) all maintain plant drying rooms filled with specimens collected locally. One of the oldest living traditionally trained Ayurvedic doctors in Nepal until his death five years ago, Siddhi Gopal Vaidya was very concerned about the decline of medicinal plants in the country, and he held many discussions with high-level officials, including former Prime Minister Bhattarai, about the country's need to accelerate conservation efforts of medicinal plants. During my visits with him over six years at his home-based clinic in Patan beginning in 1998 when he was 92 years old, Siddhi Gopal spoke about many things related to Ayurveda, always emphasizing the role that poverty plays in the destruction of biodiversity, as the market for high altitude medicinal plants had become quite lucrative in a country with limited employment opportunities outside of the familiar farming. He himself held several *ropani* of land outside of Kathmandu Valley on which his son cultivated medicinal plants. Still, alongside his sustainability concerns Siddhi Gopal was also worried about the next generation of healers. Coming from 23 generations of Ayurvedic doctors himself, he wondered if any of his sons or daughters would take up the profession with as much passion and dedication as him. One son was apprenticing with him during the day but because he had no formal Ayurvedic medical education the son would be ineligible to legitimately practice under the government's new guidelines until he was 60 years old.

Conclusion

When I first traveled to Kathmandu nearly three decades ago, there were rice and wheat fields (*khet*) scattered throughout the neighborhoods of Thamel and comprising much of what is now the popular and developed area of New Baneshwar. Now, one finds urban farming in small vegetable plots and fruit trees, with most people buying their food supplies from the thousands of little shops and vendors scattered throughout the valley, often supplementing purchased items with food supplies from village family farms. The *khet* are gone, replaced by neighborhoods of new homes, shopping centers, business plazas, schools, and all that constitutes growing urbanization in a poor country. With the cessation of rural violence and the new government in place, return migration to the villages may loosen the stranglehold on urban infrastructure. Indeed, better informed and greater rural activity may be the context in which Ayurvedic medicine continues to be practiced and to evolve, with increased awareness of how to sustainably harvest medicinal plants. Standing in the way of such beneficial contributions to the development of a fully engaged plural medical system is government action that at times seems less focused on the real needs of the Ayurvedic medical community than on developing biomedical facilities. The medical licensing statutes, for example, serve to limit future generations of traditional providers, and the lack of minimal personnel and resource development curtails progress in formal Ayurvedic medical education. At present, the state's overall inattention to one of the country's most important and popular medical systems slows the advance of health care development.

Notes

1 This chapter focuses on non-timber forest products. Nepalis also depend on timber, a heavily exploited resource throughout the twentieth century that has achieved a comeback with community-based forestry conservation.

2 Approaches to healing that are considered Ayurvedic will be briefly delineated in a later section. This chapter will not address the issue of Indian vs. Nepali Ayurvedic medicine.

3 The raising of animals and the consumption of meat also have economic, medical, and ethnic importance among Nepal's diverse populations, but are not addressed here.

4 Ayurvedic medicines may also contain minute quantities of animal materials and metals.

5 *Doṣas* are characterized by both *guṇas* and tastes, though only bile (*pitta*) and phlegm (*kapha*) are thought to have tastes. The twenty *guṇas* are heavy/light, cool/hot, unctuous or oily/rough, dull/sharp, immobile/mobile, soft/hard, non-slimy/slimy, smooth/coarse, gross/subtle, dense or solid/liquid. The six tastes are sweet, sour, salty, pungent, bitter, and astringent. Wind is rough, light, cold, subtle, mobile, non-slimy, and coarse. Bile is unctuous, sharp, hot, sour, fluid, pungent, mobile, and liquid. Phlegm is unctuous, cool, soft, sweet, immobile, and slimy. Throughout much of South Asia people commonly apply ideas associated with the qualities of the *doṣas*, such as hot, unctuous, sweet, rough, cold, and astringent to many everyday objects, most notably food, liquids, and seasons.

6 Wujastyk cautions against using the terms 'balance' and 'imbalance' when referring to Ayurvedic states of health, as these misread into ancient texts more modern concepts associated with western lifestyles (2003: xli). On the other hand, Ayurvedic studies of all kinds use such metaphors as aggravated and displaced humors when describing health. I retain this convention here.

7 In contrast to the mechanistic model of the human body found in allopathic medicine, Ayurveda presents a fluid, hydraulic body comprised of the *doṣas* and their common sense and easily-identifiable qualities of hot and cold, rough and smooth, sour and sweet, etc., found in the living and non-living world.

8 In 2002, there were 4154 allopathic hospitals, health centers, health posts, sub-health posts, and primary health centers, and 285 similar Ayurvedic facilities. There are two regional Ayurvedic hospitals, 14 zonal hospitals, 50 Ayurvedic health centers, 211 Ayurvedic clinics, nearly 40 private drug companies, 275 Ayurvedic drug dispensaries, and a semi-autonomous drug research and manufacturing institute called Singh Durbar Vaidyakhana which is over 350 years old (Nepal Development Forum 2002).

9 Interview with Mr. Bhattarai, Managing Director of HPPCL, Koteshwor, 8/9/2000.

10 The section of the Act dealing with drugs permits their prescription, preparation, and sale only if listed in published gazette notices by the Drug Distribution Agency in the Ministry of Health, and only by individuals who have received permission to do so by the Ayurvedic Council and the Department of Ayurveda. For more details of the politics and history of the full Act, see Cameron (2009a).

11 Greatly more funding is provided to allopathic medicine through WHO, UNICEF, and USAID.

3 Perceptions of forests among the Yakkha of East Nepal

Exploring the social and cultural context

Andrew Russell

Introduction

This chapter reviews the history of perceptions of environmental change among 'development practitioners' in Nepal for the past 40 years, and how this compares with the perceptions of forests among the Yakkha of Tamaphok, East Nepal. Its central argument is that to understand people's perceptions of the environment, one has to understand the social and cultural contexts which frame and give substance to these perceptions. It is based on fieldwork with the Yakkha, a relatively small Tibeto-Burman-speaking group found primarily in the middle hills of the Koshi zone of East Nepal, conducted between January 1989 and October 1990, and supplemented by a short return visit to the village in October 1996.

My fieldwork in 1989–90 was intended to look at indigenous forms of forest management in the area as a piece of applied anthropology. But when living with the Yakkha for an extended period of time I quickly became aware that forest management was only one relatively minor issue in a plethora of concerns that were being discussed then. My research ended up broadening to look at environment and culture among the Yakkha in more general terms, with the different aspects of 'environment' providing a convenient peg on which to hang the more conventional ethnographic material that comprises the bulk of my DPhil thesis.

This chapter returns to forests in environmental and post-environmental discourses, and to the social and cultural contexts which formed and informed people's knowledge and attitudes to forests and deforestation in Tamaphok at that time. In the first section, I track the Theory of Himalayan Environmental Degradation from modernist (environmental) theory to postmodernist (environmental) myth, from certainty to uncertainty. This shift has brought indigenous forms of knowledge and practice back into the frame in a more positive light, but raises questions about the status of the relationship between 'culture' and 'environment'. What does it mean to take the social and cultural context into account, and how is this exemplified in the Yakkha case study which follows? What we see is that Yakkha perceptions of their forests are shaped by a range of shifting biological, socio-political, economic, and spiritual influences. These are somewhat at odds with development-based perceptions and attitudes to the problems they face. In indigenous terms, deforestation is the rapid and patchy process caused by sudden catastrophes or the activities of a few unscrupulous individuals taking advantage of forest rights that are community-based. The third section of this chapter looks at how further research has led to the increasing certainty of uncertainty, and the implications of this for more sophisticated theorising of the relationship between 'culture' and 'environment'.

Himalayan degradation – from modernist theory to postmodern myth

Preparatory reading for research in 1989–90 was heavily influenced by the work of Thompson *et al.* (1986), first published as a series of articles in the journal *Mountain Research and Development* and subsequently as a single volume, *Uncertainty on a Himalayan Scale*. One could understand a certain disappointment on the part of the International Institute for Applied Systems Analysis (Laxenberg, Austria) for the length of time it took this team – four years from 1980 – to conclude that, in the Nepal Himalaya, it is not a case of 'what are the facts?' so much as 'what would you like the facts to be?' It seemed there was a crisis not in the Himalaya so much as among the region's researchers, a paradigm shift from certainty to uncertainty of major proportions. This section considers the changing perceptions of environmental degradation highlighted by Thompson *et al.* as a shift from a modern to a postmodern paradigm (cf. Metz, Chapter 1 in this volume).[1]

Modernist theory

Thompson *et al.* (1986), like Ives and Messerli (1989), reflected a trend questioning what might be called the 'modernist' view of environment in Nepal. This view saw an impending ecological and demographic catastrophe in the Himalaya. 'In bygone ages the Himalaya seemed eternal; man's onslaught has rendered them among the most fragile eco-systems of the earth', wrote Moddie (1981: 342). This was the 'doomsday' scenario for Nepal, in which much of the blame for the environmental changes believed to be taking place in the Himalaya was placed on the shoulders of local people. It was the backbone to Macfarlane (1976)'s application of the Malthusian co-ordinates of population and resources to the Nepalese setting. Because of population growth, many of the subsistence resources on which people in the hills of Nepal depended (e.g. fuelwood, see Eckholm 1975) were considered to be dwindling fast. At the national level, Goldstein (1983) calculated, on the basis of the 1981 census, that the population of Nepal was growing at an average rate of 2.7 per cent per year, with a doubling time, if those rates were maintained, of 27 years. In 1960, the population had already trebled since 1850 (Macfarlane 1976: 292). Macfarlane (1968) was critical of anthropologists for not according population change sufficient weight in their analyses. He argued that population had to be seen as a determinant of social and mental structures as well as being determined by them, and suggested that 'it is demographic fluctuations which underlie many of the social and economic changes in the area' (Macfarlane 1976: 8).

Poffenberger (1980) saw a direct and negative impact of population growth on agriculture and the resource situation in general. According to Seddon (1987), grain availability in the country as a whole moved from a surplus of 66,921 tonnes in 1971 to a deficit of 108,278 tonnes ten years later. Estimates of population density against food supply indicated that, from 1971 figures, only four out of 36 districts in the middle hills of Nepal were in any sense 'underpopulated'. Tuladhar (1977) noted that while the population density of hill and mountain Nepal in 1971 per unit of arable land was greater than that of Bangladesh, double and triple cropping practices were more feasible in Bangladesh and its soils generally more productive than land in most of Nepal.

This situation had geo-political implications: 'Until well into the 1980s the Nepali hill farmer was assumed to be responsible for massive deforestation, increased landsliding, soil erosion, and horrendous downstream effects through Gangetic India and Bangladesh, all the way to the Bay of Bengal' (Ives 1999: 175). Local people (primarily rural farmers) 'were perceived as ignorant within their own environment and reckless in their destruction of it' (ibid.: 175). Thus they were fair game for an onslaught of aid and development money aimed at alleviating the problems posed by what Ives called the Theory of Himalayan Environmental Degradation. This theory, Ives argued, was a latter-day myth which had to be 'challenged at all levels' (Ives 1986: 69; see also Ives and Messerli 1989). The next section examines how this has been done.

Postmodern myth

'So what's a landslide doing in the middle of a forest?' asks the caption to R.J. Fisher's photo in Deepak Gyawali's article 'Of Rabbits, Hillmen and Muddy Rivers' (1989: 30). Gurung (1982: 6) argued, contrary to Moddie, that soil erosion in Nepal was caused by the 'high energy environment' of the Himalaya, its extreme altitude, gradients, tectonic and climatic forces making the hills and mountains a 'conveyor belt' for the transport of eroded material almost regardless of the activities of people upon them. For Gurung, rather than blaming Ganges floods on Nepalese hill farmers, people downstream should consider that, were it not for 'the transport of vast quantities of sand and silt by Himalayan rivers over millions of years, there would be no Gangetic plain and no developed economies to contrast with the poverty of the hills' (Gurung 1982: 7).

In addition to querying the mechanics of soil erosion in Nepal, the state of Nepal's forests was also being reassessed. Macfarlane (2001) could see little evidence of forest decline around Thak over a 20-year period. Thompson *et al.* (1986) cited an observer who had found that large tracts of the forests around Khumbu had remained intact over the years. Observations like this lent credence to Gilmour *et al.*'s assertion that the crisis dimensions of deforestation in Nepal had often been overplayed (1989). Mahat *et al.* (1986a, 1986b) challenged many other preconceptions about deforestation in the country. They considered it to be a far from recent phenomenon (thus displacing its assumed link with population growth) and suggested that it might in fact be decreasing, influenced as much by government land-use policies and land taxes as with the demands of subsistence agriculture for more land (cf. Bajracharya 1983). Rather than local people being the mindless instigators of denudation, indigenous forest management systems had been found to be operating in many parts of the country (e.g. Cronin 1979; Hawkins 1981; Chapagain 1984; Fisher *et al.* 1989). Interest in such activities was stimulated by the initiation of some ground-breaking legislation concerning 'community forestry' by the Nepalese government in 1978 (see Nightingale, Chapter 4 in this volume).

To support their argument that no one really knows the full situation, Thompson *et al.* (1986) pointed out how expert estimates they collected of such relevant variables as 'per capita fuelwood consumption' and 'sustainable yield from forest production' varied by factors of 67 and 150 respectively. They argued that 'in coming to grips with uncertainty', different methodological approaches were appropriate. They saw a need to shift the 'locus of expertise' away from those whom they call 'centrist grand designers', able to work only with simple problems, to 'fringy tinkerers', better able to handle dynamic systems (ibid.: 150), and to incorporate greater portions of what they call the 'natural:cultural template' into their analyses. In other words, bring on the anthropologists!

Environment and culture

However, in bringing on the anthropologists, there is a risk that they themselves may carry certain assumptions concerning the role of local knowledge and practice.

If the dire prognostications of the doomsday scenario writers have not come to pass, for whatever reasons, then it must be because the hill farmers, so long disparaged and treated as 'part of the problem', were doing something right. Several studies identified a wealth of indigenous knowledge regarding environmental processes such as landslide risk (Johnson *et al.* 1982). Fricke (1986) saw population expansion as a prime mover in people's 'adaptive strategies' in Timling, a Tamang village in central Nepal. Dahal (1983) similarly recorded a number of processes he considered responses to population pressure in a village in Ilam district, East Nepal. These include the intensification of agriculture, the adoption of cash crops, the increased use of marginal lands, the generation of other sources of income (for example, through the development of trade or by joining the army), out-migration, and fertility controls. Thus, while not denying the importance of population pressure on social life, researchers were coming to see such responses as more positive and dynamic in their effects than had previously been imagined.

According to Vandana Shiva, forests throughout South Asia are viewed by the people that use them as 'the highest expression of the earth's fertility and productivity' (1989: 56). She cites Myers' statement that traditional perceptions of forests 'convey a sense of intimate harmony, with people and forests equal occupants of a communal habitat, a primary source of congruity between man and nature' (ibid.). However, the tendency to assume that environmental relations in more technologically simple societies are generally adaptive is based on ideology rather than necessity. Members of some societies do appear to live in a degree of adaptive harmony with their environment – linguistically, behaviourally and ideologically. Such are the Mbuti pygmies, for whom the Ituri forest 'is their world, and in return for their affection and trust it supplies them with all their needs' (Turnbull 1961: 14; cf. Mosko 1987), in contrast with the BaBira villagers, for whom 'the forest was hostile, something to be feared and fought' (Turnbull 1961: 171). The Mbuti relationship with their forest, though, is based on cultural ideology which may or may not be similar (or coincide with ecological necessity) in other populations. Other forest-dwelling groups, such as the Semang of Northern Malaysia described by Rambo (1985), may have a less benign relationship with their environment. Geertz similarly cites Freeman's example of the Iban of Borneo, the self-ascribed 'eaters of woods' (1963). Even Rappaport, who presents the Tsembaga as managing their ecosystem in an extremely rational manner, admits that in other societies, 'maladaptive, dysfunctional, or contradictory equilibration may not be at all uncommon' (1979: 100). It is unnecessary to assume that, just because it exists, a society is in equilibrium with its environment, or will remain so in the future.

The issue of 'adaptive'/'maladaptive' is part of a larger discussion within anthropology and allied disciplines concerning the relationship between culture and environment. According to Visvader, 'in a relatively unchanging world the form of culture comes to fit the form of its environment on a very basic level' (1986: 126). This is queried in the case of Nepal by Schroeder, who argues that 'agricultural characteristics are strongly influenced by local agronomic and ecological factors and do not respect ethnic or cultural boundaries' (1985: 35). Perhaps this is simply based on ecological commonsense. According to Johnson:

> In the ecological sphere, we should expect cultural notions and actual behavior to be in close correspondence, since the validity of cultural rules is constantly being tested in environmental interactions where the price of confusion and error may be malnutrition, serious injury, or death.
> (1974: 98)

However, if people's views on specific features of their environment such as forests are pandemic (Shiva) or their practice with regard to agriculture so determined by environmental factors (Schroeder; Johnson), is there any point in studying the people:environment perceptions and relationships of a group like the Yakkha at all? To do so requires taking the social and cultural context more seriously than the theoreticians at either end of the 'culture–environment' spectrum would allow. Unfortunately, doing this takes longer than the simple (and often simplistic) study of a specific feature of a cultural group. Müller-Böker, in her study of forest knowledge among the Tharu of the Tarai, concludes:

> Knowledge about nature is closely related to the utilization of natural resources. The evaluation and perception of natural environment, however, are greatly influenced by cultural and religiously-toned concepts not directly comprehensible to an outsider . . . Why are forests highly revered among the Tharus as the domain of gods and spirits, but not among the Pahariyas? Even if our answers can be only speculative, it is at least worth observing and finding out these local conceptions, because they influence the behavior of the people towards nature and natural resources.
>
> (1991: 112)

The second part of this chapter looks at such factors in the lives of the Yakkha of Tamaphok.

The social and cultural context of forests in Tamaphok

Tamaphok is in many ways a heartland of Yakkha language and culture, a dispersed settlement situated on a north-facing slope of the Maya Khola valley at an elevation of between 4000 and 7000 feet. The views northwards from Tamaphok to Mount Makalu and some of the other snow-capped Himalayan peaks are often stunning, although seldom remarked upon by village inhabitants. In addition to my fieldwork in 1989–90 and return visit in October 1996, I have maintained contact with people from Tamaphok through Yakkha serving in Gurkha regiments stationed in the UK, the visit of the village 'sister' Kamala in July/August 1993 (Kohn 1995), and through letters and emails from Kamala and other Tamaphok friends. My original research was funded by a UK Economic and Social Research Council 'CASS' award designed to foster collaboration between academic and non-academic institutions, in my case, the University of Oxford and the Overseas Development Administration (ODA, now the Department for International Development). I selected the Yakkha as an ethnic group that, to the best of my knowledge, had never been studied before, who appeared to be located in the centre of the ODA's Koshi Hills Area Rural Development Project (KHARDEP). The intention was to study indigenous forest knowledge and management systems among the Yakkha as a contribution to the Community Forestry Project that was underway within KHARDEP.

The postmodern turn in environmental anthropology (indeed, social anthropology in general) meant I was sceptical of large-scale quantitative research techniques (which would take a lot of time to find out a lot about a little), preferring in-depth qualitative research methods (which would take an equal amount of time, but would find out a little about a lot more). I stayed in Tamaphok for 15 months in total; the first seven months living in Budhilal Linkha's house with his daughter Kamala and his wife, Ama, the rest in a house further up the valley that had been abandoned by a family who had migrated to the lowland Tarai. These habitations provided the immediate social contexts for participant-observation research. I also undertook a livelihoods survey of 20 households in different parts of the village. These were selected at random by a research assistant to provide a diversity of socio-economic backgrounds. Periods in the village were interspersed with short breaks to the Tarai or Kathmandu, longer in the time of the pro-democracy movement disturbances (which reached their height during a visit to Kathmandu in April 1990) and when I broke my arm (due to an altercation with the spirit soghɛk[2] on a slippery forest path above the village in March 1990) (Russell 2000). All this was underpinned and dependent upon the continuous and intensive learning of two languages. Nepali, the *lingua franca* of Nepal, is understood and used by everyone in Tamaphok, and is the language I decided to use for most of my research. The Yakkha language had been only scantily recorded previously. The German linguist Alfons Weidert had done a study, but he had been robbed on a train in India and all his records were lost. Yakkha was understood and used by the majority of Yakkha in Tamaphok, and was clearly a key to understanding important aspects of Yakkha identity as well as being a marker of Yakkha identity in its own right.

In my research, I followed the lead of local people in terms of what seemed important or significant in their lives. Forest knowledge became less of a primary focus as I became better attuned to the social, cultural and environmental features which excited the Yakkha themselves. Forests did not figure prominently in this regard, any more than mountains did, and to follow such a 'specific feature of a cultural group' agenda would have been an injustice to the multiplicity of features and the diversity of cultural forms among the people. So, while I kept my eyes and ears open to activities and/or discussions concerning the forest, what follows is the product of a more piecemeal (and hence postmodern?) approach to ethnographic research, but one which I feel better reflects the disparate and fragmented ways in which 'forest knowledge' is contextualized among the Yakkha (cf. Campbell, Chapter 2 in this volume).

Landscape and maps

Sagant (1976) could have been writing about the Maya Khola valley in describing the typical scene of the Mewa Khola valley in the predominantly

Limbu district of Ilam: a swathe of houses and fields across the slopes, broken only by thickets of bamboo and groves of trees, often between two wooded belts, one by the river and one by the crest. These extremities were distinctive. Rivers were often malarial, rocky and gorge-like, and both they and the forests belts were wild areas inspiring fear (ibid.: 32). Like the Limbu, for many Yakkha the riverine and ridge-crest forests were wild, jungly places, the haunt of ghosts, spirits and wild animals. Sagant's description encapsulates both the physical and symbolic/emotional aspects of the people:environment relationship I wish to highlight in this section.

One day Budhilal Linkha, who was *pradhān pañc* [panchayat leader] at the time, produced a map he had drawn of the *pañcāyat* [the unit of local government preceding the pro-democracy movement]. It is oriented on the page according to topography, with the highest parts of the *pañcāyat* (towards the south) at the top, and the lowest riverine areas (towards the north) at the bottom. It shows the key features of Tamaphok as he saw them – the schools in the different wards, and the *pañcāyat* office next to Sri Chaumunde Secondary School, Tamaphok, ward no. 6. It shows the footpaths linking different parts of the *pañcāyat*, and marks and names many of the streams running through the four major watersheds it traverses. There are no forests, though. The only distinction of land type he makes is *kharka jaggā* [open or pasture land – this would also include much of the forested areas of the upper slopes]. Under the *kipat* system of land tenure, land was collectively owned and was distributed among clan members under the jurisdiction of the *majhiyā*, traditional leaders appropriated by the invading Prithvi Narayan Shah to act as intermediaries between the local people and the state. With the demise of the *kipat* system, bolstered by the arrival of the cadastral survey to apportion private title to land during the time of my fieldwork in 1989, the role of the *majhiyā* had become largely symbolic. They were expected to collect a token amount of land tax [*tiro*] from household heads once a year, which they took to the district headquarters in Khandbari. They expected payment in *raksi* [spirits] and 5 Nepali rupees (about 10 UK pence or 15 US cents) from a bridegroom taking one of their kinsmen's daughters away in marriage, and it was their role to perform the main sacrifice to Durga at the *Dasaî* festival each year. The forests of the upper and lower slopes were nationalised by the government by two main Acts, in 1957 and 1961, but the forest guards based in Chainpur were far away and these tracts of forest were still regarded as communally owned, reflected in the term '*hāmro ban*' [our forests].

It would be wrong to make too much of the absence of forests from the *pradhān pañc*'s map, but it is tempting to see it as a reflection of the relative insignificance of forests in the everyday conversations of the Yakkha. Even though people were reliant on trees for firewood, fodder, wood for building and many other purposes, compared to the more southerly *pañcāyat*s in the districts where KHARDEP staff were concentrating their activities, Tamaphok still had relatively abundant forest reserves and so concerns about forest resources were not high on people's list of priority problems. Moreover, the development of cardamom as a cash crop among wealthier landholders in Tamaphok had stimulated the planting of new trees to meet the shade requirements of the crop. There was little evidence, in the six years between my first and second (short) fieldwork period, of serious depletion of forests, although there were some cleared tracts of forest land at the highest elevations that were said to have been made by individuals trying to establish land claims prior to construction of a road from Basantapur to Chainpur (and onwards up the Arun valley to the 'Arun 3' dam project, subsequently cancelled). Clearly the concept of '*hāmro ban*' did not impinge upon such actions.

The collection of forest resources (predominantly firewood) was just one of the many hard subsistence activities that people had to do, and was woven into the agricultural calendar like everything else.[3] Carrying firewood home from the forest in a *ḍoko* [basket], a strap taking the weight on one's forehead and making one's neck and back ache, was regarded as work for women and children. It was unusual to see a man carrying wood back to his house. Of their many subsistence activities, this was the task most women said they liked the least. A full basket of wood could weigh anything up to 50 kilograms. The over-riding (and pan-Nepal) view of all these activities was one of drudgery and a sense of suffering, pain and hardship encapsulated in the Nepali word *dukha* (pronounced <u>tukha</u> in Yakkha, and often tagged alliteratively with the word for village to make '<u>tukhi ten</u>' [village of suffering]). Maybe there was some relationship here with the classic Hindu *yuga* doctrine that life-spans were decreasing. A Yakkha man once reflected on how, in Krishna's time, men used to live 300 years. Now, with the production and consumption of crops such as rice, millet and maize, they could live no longer than 110. Subsistence living was therefore something of a curse.

Legends, spirits, and emotions

It might seem odd for Sagant to cast the forest, an integral part of the farming system, in the role of cultural bogeyman. However, while the forests were known as a source of wood, medicines, strawberries and chestnuts, in Tamaphok the acknowledged importance of forest products did not detract from people's fear of the places they came from and the disdain with which they regarded the activities that had to go on in them. One man remembered the days when carrying goods from Dharan to Taplejung could earn him 120 Nepali rupees (worth a lot more in those days); his main fear along the route was the terrifying forest between Basantapur and Chauki (a hamlet at the head of the Maya Khola valley). There was at that time no human habitation between these two places.

Two young women carrying 40kg loads to the Tamaphok shop hurried up as I approached the hill top forests on the path from Basantapur one day. 'Can we walk with you?' they asked, 'a *lāhure* [retired serviceman] died here once.' They were also being troubled by a leering drunken man, but their fear of ghosts seemed equally genuine. Not all Yakkha men and women admitted to being reluctant to walk through the forest alone, but it was interesting that, when striding out to Basantapur from Tamaphok in a group, it was generally not until we were out of the forest and into the *lekh* [highland, mountain] pastures beyond that my village companions would take their leave of me (if they did), even if my pace prior to this may have been unbearably slow for them!

The somewhat troubled relationship of the Yakkha to their forests is reflected in the myths and legends described in various parts of their muntum [ancestral lore, orally transmitted]. One legend concerning the origin of the placenames in the Maya Khola valley describes a group of Yakkha running as fugitives from the local Limbu inhabitants of Pokhlabhung in neighbouring Tehrathum district, having set fire to the forest there while trying to 'smoke out' a porcupine from its lair:

> The Limbu caught up with them at Chamlung and two or three were killed. However, one man, Lalubhang, escaped. He was a strong man and ran off carrying his younger sister on his bicep. They reached Mamling [a village further down the valley from Tamaphok] where Lalubhang was hidden in a house so that the Limbus could not find him. 'You must be my servant,' said the man who hid him, and made him a *nokar* [servant]. Lalubhang was sad to be a servant. 'Why should I be a servant? I am also a Yakkha,' he thought. He had been a *dhāmi* [spiritual healer/seer] before. He decided to use his powers to send his master's daughter to the jungle. When this happened, his master was very sad. He called on the powers of many *dhāmi* to find his daughter, but they could not. He asked far and wide for *dhāmi*s. Lalubhang's younger sister told him that her brother was one. When approached, Lalubhang gave a thread to his master and after doing *henemene* [augury] told him to follow the thread to his daughter. She was high up on a cliff in a cave under the rock. So Lalubhang cut a tree to make a ladder and went up. He gave her birds' feathers as a cure for dumbness. Her father was very happy, and gave her to Lalubhang in marriage. Lalubhang gave the father his younger sister. Thus he became the same as his master, and lost his servant status. Thus commenced the Huturai clan [from the Yakkha hu'ma; Nepali *ghusrinu* – 'to be a fugitive'].

This oral history of the Huturai Yakkha having 'done wrong' by the forest and become fugitives because of it is matched by a story about the origins of a spirit associated with the Linkhā clan that was caused by a hunting transgression:

> Linkhāsam came from a branch of the Linkhā clan called the Iknap. This group of Linkhās lived in Mamling but used to come to Kuntang [between Mamling and Tamaphok] to hunt. When there, they killed a

> deer. According to their custom, when a deer or goat was killed, the cloven hoof should be cut in two through the cleft, but on this occasion they failed to do this. Therefore the *deutā* [clan deity] became angry. The culprits became ill and died. From them dying, the Linkha sakmabu arose.
>
> After this, the Iknaps were chased out of Mamling, and came to Tamaphok where Challas [another clan] were already living. They went to Panglum [an area slightly down the valley and below the main part of Tamaphok today; also called Deurali]. From there they went up to Otemmatol hunting. Their names were Jitika, Jakhiba, Maghan Singh, Jugut Ram, Hengware, Khengwa Ram, Nabuk Sām and Thebuk Sām. They cleared the jungle at Panglum. An elephant and a jungle buffalo used to live at Otemmatol near a big pond. They went hunting there and killed the elephant, cleared the jungle and made a field. They drained the pond [otemma: flat area from where water has been drained].
>
> They built houses there. After that Nabuk Sām and Thebuk Sām died while hunting. From the two of them came the cyang [spirit] Linkhāsam. Because of them, when the Iknaps kill an animal while hunting they must kill a cock as *pūjā* to Linkhāsam and offer the blood.

Thus many troublesome spirits were said to live outside the normal confines of social life, in the forests and other wild places. One such spirit was minuma cyang [cat spirit], whose mewing would lure an unsuspecting traveller off the path to their death. Tamburang (cf. Sagant 1969) and Jangali were other spirits associated with the forest. Then there were the spirits associated with 'bad' or 'untimely' deaths [*akal mṛtyu*] such as meca cyang, which were said to be the spirits of dead babies, and soghɛk, spirits of people who have died in accidents which was the spirit said to have afflicted me when I fell and broke my arm on the path leading down into the village. Judging from the story I was told concerning soghɛk, it was lucky I did not have a more serious accident:

> Seven coloured rays from the sun came to earth and dried up everything. Only there was one porcupine which they could not dry, because of its spines [uthing]. The porcupine shot its spines at the sun, but they were not long enough to reach it. Instead, as he shot them, they joined up and reached there.
>
> On the earth they became bamboo. Many bamboos grew and reached the sky. The sun could not dry the plants. Then a small mouse hid in the bamboo, and was not killed, although its back was reddened by the sun. From inside the bamboo it climbed up through the joints by nibbling through them. It came out at the top and bit the sun's rays. The wind blew the bamboo back and forth and the mouse bit off all seven rays. And where the mouse bit the rays, blood fell to earth. Wherever it fell, on rocks, on trees or on earth, the blood did not dry. People going by these

places fell over and died. They had touched soghɛk and became soghɛk themselves.

Wherever soghɛk plays and men pass by, soghɛk will touch them and horns form. Only *dhāmi*s can see these horns. Within 15 days, if not treated, this person will have an accident. Only certain Rais have the power to kill soghɛk. This is because Rais had a fight with soghɛk and made it run away. It hid in a hole in a tree and the Utangi/Batangi Rais came and covered it over thinking they would kill it. But soghɛk knocked over the tree and escaped. The Rais returned and pursued it. Soghɛk told them 'Don't chase me, I won't do anything to you. If you kill me another soghɛk will come in the form of a bat.' So even though a Rai can kill a soghɛk, another will come and people still have accidents. Yakkha cannot kill soghɛk, but can control its effects by burning off its horns.[4]

Wild animals

As well as spirits, forests were also regarded as the abode of wild animals, even though many of these were rarely, if ever, seen. A common term of rebuke for young children was *jaṅgali mānche* [jungle man], which indicated a person who was wild, uncivilised and lacking in education. The only wild animal for which any degree of warmth was felt appeared to be the mongoose, which controlled rats and mice in the millet or maize fields. For the rest, forest animals were generally abhorred because of the damage they did to crops or because of their predation on domestic animals such as ducks, chickens, and young goats. The list of animals negatively associated with the forest in some way included tigers, bears, wild boars, deer, eagles, jackals, wild cats, monkeys, porcupines and leeches.

Of this list, only jackals, cats, monkeys and leeches were common, and of these jackals were probably the most troublesome in terms of their predations on domestic food supplies. At night, at any time of year, it was not uncommon to be surprised by the piercing, eerily human cries of these animals, which hunted individually but communicated within their group, echoing up and down the valley. As the cries became louder, people sprang into action from their beds or sleeping mats, meeting these wild transgressors with the domestic sounds of banging lids and a distinctive, yodelling cry; too late for many a duck or chicken, however. Of the five ducks owned by Budhilal Linkha's family, for example, only one actually reached the mouths of its owners (and their guests) during the year that I lived in their company. During the summer, moreover, jackals were said to eat the maize cobs growing in the fields.

Monkeys were also a serious threat to the maize crop, as well as being the scourge of cardamom growers. The guns to be heard going off in October/November were generally aimed at monkeys in the cardamom groves. Various comments about people revolved round their supposed similarity with monkeys, in appearance or behaviour. When I ate my maize off the cob with my teeth, rather than using the side of my thumb to take off the kernels a line at a time, for example, I was told I was eating maize like a monkey. I was also persuaded to shave my beard off while in the village, not only because I was said to be scaring the children but also because I looked like a monkey!

All the other animals were rare. For example, the last tiger in living memory had been killed over 30 years before. It had come out of the forest one morning, and had carried off a prominent villager's dog. The beast was tracked by a trail of blood to a tree in the forest above the (current) school where, as evening fell, a posse of men gathered and shot it. Some people suggested that jackals had taken the place of tigers (an indigenous sense of ecological 'niche') since no jackals were said to be found where tigers lived. Wild boars had not been seen for 13 or 14 years. Bears were said still to be in existence in the forests, but no one reported seeing any during my fieldwork. Nor did anyone see a porcupine [yakpuca], although Budhilal Linkha kept a porcupine quill as a memento and claimed there were still some to be found in the forest. They were remembered fondly as providing the most delectable meat. Their perceived danger to the maize crop was reflected in a verse said to have been made up by a Damai-caste tailor in Phumling to entertain the children: 'Eko, hici, sumci, *makai khāne dumsī*' ['one, two, three, maize eating porcupine' – using the Yakkha numbers but Nepali name for the animal].

The situation was rather like that of the Gurung village of Thak described by Macfarlane:

> The forest is no longer within easy reach of the village, and one has to travel many miles before reaching areas where most of the game has not been exterminated. One or two villagers were still keen huntsmen, but despite many expeditions only one small deer was shot during our fifteen months in the village.
>
> (Macfarlane 1976: 30)

Despite the disappearance of many wild animals, they lived on in the imaginations of the Tamaphok Yakkha and, I would suggest, were another potent factor in how people viewed their forests. Unlike the Gurung whom Macfarlane reports had dispensed with their bows (ibid.: 30), Yakkha men in Tamaphok still often walked around the village carrying their pellet bows (cigik, *guleli*) and a pocket full of kambruk (*māṭeṅrā* – clay pellets), although now their weapons were only likely to be used to shoot at birds or errant dogs. Skill at making them was highly regarded. I never saw a member of any other caste or ethnic group in Tamaphok carrying a bow.

Perceptions and actions

These predatory creatures, as well as spirits such as soghɛk, provided potent physical and metaphysical reasons for perceiving forests as less than benign environments. In cutting down forest trees for fuelwood (as one had either to do oneself or arrange for others to do at least once a year), there was a sense that one was not only contributing to the depletion of forest cover but was cutting back on the abode of eagles and wild beasts of all descriptions. While people were dependent on forest resources for their fuelwood, fodder and many other subsidiary products, there was nonetheless an ambivalence in people's remarks about the decline in forest resources over the previous decades. Their practices, based on need for the products they provided, acknowledged that dangerous elements in the forest environment could harm them, and sought to minimise this risk. If there was a concern for protecting essential resources such as firewood and fodder, it was balanced by a competing concern to eradicate the dangers that lurked within forests.

This ambivalence may have been linked to people's perceptions of the causes of forest depletion. Many suggested that natural hazards, such as fire and landslides, rather than human agency, were the major factors in forest decline (hence giving indigenous support to Ives' critique of the Theory of Himalayan Environmental Degradation). It was interesting that when I asked people for examples of where forest cover had diminished significantly over the years, the extensive area of forest which had previously existed near the school was most frequently mentioned. This had been destroyed by a fire ten years previously, and having been close to many houses, had been much used before its sudden destruction.

The certainty of uncertainty

Much subsequent work has gone on to prove Thompson *et al.* were broadly correct in their interpretation of the situation. In a hill village in central Nepal researched over a ten-year period Fox (1993) found that, despite an annual population growth rate of 2.5 per cent, the forests were in better condition in 1990 than they were in 1980. This situation was confirmed in the Eastern hills by Virgo and Subba (1994), which they largely attributed to the increase in numbers of fodder trees on *bāri* land. In the case of Tamaphok, as we have seen above, a major factor has been the propagation of shade trees to protect the new cardamom crop. These trees can also be used for fuelwood. Macfarlane (2001) has also rescinded some of the gloomy prognostications he made in 1976, although he emphasises that people are poorer than they were before and the future for the Gurungs of Thak is far from rosy.

> If anything, the forest above Thak is in better condition than it was in 1969. The tree cover is growing back nearer to the village. This is the result of a slight decline in the need for firewood and also because of another large change . . . the dramatic decrease in number of larger animals. So, although there has been erosion and loss of some land through landslides, the catastrophe which I predicted in relation to the forests has not occurred.
>
> (Macfarlane 2001: 106–7)

The major variable that Macfarlane failed to predict in 1976 was migration, i.e. the ability of people to step outside their farming ecosystem altogether. Migration was a topic that even in 1989/90 had started to become more prominent in the literature (e.g. Thapa 1989) and was an important aspect of life in Tamaphok (Russell 1992). Whatever its long-term effects on the receiving areas, migration was important not only as a 'safety valve' for the growth of population in the Nepalese hills, but also as a means of ensuring that people did not have to deplete their environment in the name of subsistence. It suggested to me that the most significant cultural differences in terms of people:environment relations in Nepal was not so much in their response to *dukha* [suffering], which was universally shared, but in their attempts to convert *dukha* into *sukha* [prosperity; well-being] (cf. Fisher 1987).

The uncertainty about what was actually happening has stimulated more careful, empirical studies of local processes. Wu and Thornes (1995) have demonstrated that terracing on Nepalese hill slopes has generally positive effects. Gerrard and Gardner (2002), in a study of landslide events in the Likhu Khola basin (north-west of Kathmandu), found that most slope failures were on irrigated *khet* land and were small and easily repaired (although with labour cost implications), and that deforestation would not necessarily lead to land degradation. More attention has been paid to the use of trees planted on private land (Carter 1992), on individual crops such as the Nepal pepper (**Zanthoxylum armatum;** *timur*) which is highly valued in India for use in health care products (den Hertog and Wiersum 2000), and on the trade in medicinal and other forest products (Edwards 1996). More is understood about the importance for biodiversity of the remnant stands of riverine forest as 'riparian corridors' for subtropical species and an important resource for local people (Zomer *et al.* 2001). Such studies support many of the observations made about forest cover and levels of environmental degradation in Tamaphok, East Nepal, and help to put other observations, which did not appear so significant at the time, into high relief.

Further progress has been made on refining the measures by which assessments may be made. Gautam and Watanabe (2004), for example, use a comparison of map analysis, air photographs and enquiries of local people to query the reliability of land use maps in the Kanchenjunga Conservation Area, which they suggest have tended to overestimate the extent of forest depletion by as much as a factor of 2. This underscores the suggestion by Thompson *et al.* that it is not so much 'what are the facts?' as 'what do you want the facts to be?'

As with all paradigm shifts, there have also been qualifications made to the extreme form of postmodernism represented by Thompson *et al.*'s work. Jodha (1995), for example, tries to reconcile the opposed views that environmental change is caused by people and that it is caused by natural processes. Blaikie calls the move from 'what are the facts?' to 'what would you like the facts to be?' 'an extreme response to uncertainty and one that tends to be wasteful of what we can be *reasonably* certain' (1988: 139). Messerschmidt

(1987: 374) concludes, 'Though I tend to sympathize with much of their perspective, I am nonetheless convinced that environmental problems do exist – in the minds of scientists and developers and, not the least, in the perceptions of the Himalayan folk themselves'. Jackson *et al.* (1998) support Gilmour (1988)'s conclusions that local farmers were taking the initiative in reforestation rather than doing so as part of institutionally-funded programmes. However, they still found evidence of upper forest slope degradation due to population pressure and lack of well-organized land management policies and practices. As Kollmair and Müller-Böker argue, 'After many years spent in the scientific deconstruction of the Theory of Himalayan Environmental Degradation, we should remember that things (and forests) could also deteriorate in reality!' (2002: 81).

Culture and environment: contextualizing the relationship

The material presented here represents a qualification of the assumption that indigenous knowledge and practice with regard to 'the environment' is necessarily adaptive or conservationist. Shiva's description of people's regard for forests in South Asia simply did not apply to the lived experience of Yakkha in Tamaphok. This would appear to support Hallpike (1986) who says that, in most cases, culture and society are neither adaptive nor maladaptive, but simply non-adaptive. It might seem strange that the Yakkha exhibit the attitudes they do, particularly since, before the introduction of rice cultivation, they had undoubtedly been more dependent on the forests for sustenance than was the case in the late twentieth century. Prior to rice and settled agriculture, the Yakkha were most probably slash-and-burn agriculturalists. Maybe this background explains the observation made to me by a Brahmin schoolteacher in 1996 that the Yakkha were better at horticulture than Brahmins and Chhetris (who had migrated into the area over the past 250 years), who were more skilled in pastoral work with cattle and goats (as Subba describes for the Limbu, Chapter 6 in this volume). However, such echoes from history were ephemeral indeed, and the over-riding impression from the Yakkha case is to support Schroeder's view (1985) that agricultural characteristics do not respect cultural boundaries in any general way.

While some shifting cultivators exhibit a sense of harmony with their forest environment, it would seem that, rather than an adaptive necessity, such ideologies operate independently of the ecological basis of the societies in which they are found (cf. Rambo 1985). It is also possible for ideologies to contradict behaviour, either because of stronger competing ideologies, or for economic and political reasons outside of actors' control (cf. Bennett (1976: 141–4) for Japan). As Low and Tremayne (2001: 16) put it, '[P]overty, political insecurity, economic systems and population pressure all play a part in the way that people refigure their sense of the world and their interaction with the environment.' In other words, perhaps the Yakkha had neither the time nor the resources to look on the forest in a romantic light. For them, ideology (that they were dangerous places) and practical necessity (to cut them down) seemed to co-exist in an uneasy harmony. True, people are a part of their environment, but, as Bennett (1976: 193) puts it, '[S]ocial systems are organized differently than biological systems, and no amount of analogy-building will alter this fact or make social systems more like biological.' In other words, the two are separate and may operate conterminously but with only contingent effects one upon the other.

Conclusion

This study shows that people are part of more than a merely subsistence environment. In order to appreciate the social and cultural context of people's environmental relations, we need to incorporate a wider frame of reference into our analyses. This frame may encompass elements (such as the spirit world) that would not figure in an environmentalist approach to people:environment relations, and would serve to increase the complexity of what I have called the post-environmentalist approach (through the incorporation, say, of legends and emotions). We need to step outside the ecosystem of development discourse (cf. Campbell, Chapter 2 in this volume), just as local people may step outside it through migration.

The postmodern turn in environmental studies and anthropology has had its effect in debunking the theory of environmental degradation in the Himalayan region and turning it into a modern-day 'myth'. It has led to more careful analysis of environmental change in Nepal, and the role of local populations in knowing about and managing their environments. However, in bringing local knowledge back into the picture, there is a danger in assuming a second, romantic myth about indigenous people, namely that their views with regard to forests and forest products are going to be positive and conservationist. The question of how environment is perceived by a group such as the Yakkha must incorporate broader issues of society and culture into its answer.

Notes

1 The tenor of Thompson *et al.* (1986), which I call a postmodern environmental anthropology, has interesting parallels with the so-called 'postmodernist' trend in contemporary social anthropology. To quote Marcus and Fischer:

> The authority of 'grand theory' styles seems suspended for the moment in favor of a close consideration of such issues as contextuality, the meaning of social life to those who enact it, and the explanations of exceptions and indeterminates rather than regularities in the phenomena observed – all issues that make problematic what were taken for granted as facts or certainties on which the validity of paradigms had rested.
>
> (Marcus and Fischer 1986: 8)

2 Throughout this chapter, Yakkha words and phrases are underlined, Nepali words and phrases are in italics, and Latin names are in bold.

3 Most wood cutting and carrying took place during the winter and spring months. People avoided going to the forests during the monsoon because of the leeches, the wetness of the wood and their heavy agricultural workload. The bulk of the cutting took place between mid-February and mid-April, and wood was then carried to people's houses from mid-March to mid-May. Only if the supply was proving insufficient for the whole year would people make further sorties to collect baskets of wood after the monsoon, in October or November.

Various trees were favoured for firewood. Among the most popular was the Nepalese alder [**Alnus nepalensis**; saksingbung – cɛkcibu in old/Ankhibuhi Yakkha; *utis*], an oak species [**Quercus lanceaefolia**; *pātle*], the evergreen **Eurya acuminata** [*jhĩnganī*], a laurel [**Litsea puleherrima**; *phusure*], Blue Japanese oak [**Quercus glauca**; *phalāt*], **Symplocos spp.** [*kholme*], and pommelo [**Citrus grandis**; *bhogaṭe*]. A tree called *ghangā~rū* [**Crataegus crenulata**] was cut but was less liked because it had prickly branches. Wax tree [**Rhus succedanea**; kag bhalag; *kāg bhalāyo*], when green, was said to have had a resin which could cause nasty weals to those susceptible to them which it was said could be treated by a *dhāmī*'s mantra. While kag bhalag was usable as firewood, it was liable to explode in the fire, a property which was put to symbolic use by some *dhāmi*s in the course of their healing rituals.

Other forest products were valued. Nepal chestnut [**Castanopsis indica**; pohẽ'mu'; *dhāle kaṭus*] was favoured as the leaf to fill with *surtī* [raw tobacco] and smoke. Its leaves were also valued as fodder. The forests also provided food, if need be, in the 'hungry season' early in the Nepalese year. A local *kāmi* [blacksmith caste] family interviewed could remember gathering the shoots of pigweed [*laṭṭe sāg*] when food was short. Another crop which was abundant in the forests during June was wild strawberries, but nobody seemed to bother collecting these small but delicious fruits. Horse chestnuts [**Aesculus indica**; hipsuwa; *masuri kaṭus*] with edible fruits larger than those of *dhāle kaṭus* were available in the autumn. Gathering them for sale appeared to be the labour of the poor and indigent. Medicinal plants [*jaribuṭī*] from the forest were also important. Just about everyone knew about the healing properties of **Swertia angustifolia** [ekhen; *ciraito*], for example, the bitter leaves of which were used to treat fever.

4 Pokhrel *et al.* (2040 V.S.: 1375–76) describe a spirit '*sogha*' as "in Limbu culture, a spirit believed to hunt people for an untimely death". Thus soghɛk would not appear to be a spirit unique to the Yakkha or Rai pantheon.

4 A forest community or community forestry?

Beliefs, meanings and nature in North-western Nepal

Andrea Nightingale

Introduction

In this chapter, I highlight how the history of community forestry in Nepal has been embedded within different understandings of ecosystems and development, including the Theory of Himalayan Environmental Degradation (THED). I briefly review the national programme and then examine the daily functioning of a user-group in Mugu District, north-western Nepal to give a nuanced snapshot of community forestry in action. Through this narrative I draw out the ways in which different knowledges are employed by different actors within community forestry contexts and how the methodologies and theoretical constructs used to investigate the program produce different knowledges of community forestry and the forests themselves.

The history of management in Mugu is extremely important for understanding current practices, ecologies and struggles over the forest. I tell a history of forest use that is compiled from oral history accounts and historical documents. This narrative illustrates the importance of forests throughout the twentieth century and helps to explain current interpretations of community forestry. In my narrative, I highlight details from the oral histories that illustrate the boundaries of the forest for the tellers; these boundaries are not simply spatial or ecological, but rather demarcate a variety of social and cultural aspects of the forest.

In this chapter, I move between a historical materialist interpretation of socio-ecological change and a relational one. In other words, I show how the biophysical forest is mutually constitutive of the social politics of forest management (Peet and Watts 2004), but also I explain how it is that a 'forest' has come to matter in the first place. Studies of community forestry are predicated on a *forest* and yet a relational perspective demands that I explain the processes that produce a 'forest', materially, symbolically, and politically (Callon 1986; Latour 1997). How is it that community forests are sustained as 'forests'? The narrative here thus shifts between a history of how forests have come to hold the significance they have, and an explanation of the processes that produce a forest as a culturally meaningful category (see also Fortier, Chapter 5 in this volume). It is not simply the details of management, ecological conditions and contestations over knowledge that are significant, but also how these aspects of forests are produced, framed, (re)told and therefore brought into view (Braun 2002). The ecological processes of water and soil movement, seedling establishment and tree growth are only sustained in that configuration by harvesting practices, management strategies and political relationships. One does not pre-exist the other and cannot be understood in isolation (Haraway 1997; Latour 1997). This kind of relational perspective thus rejects entirely the idea of 'impacts' or 'feedbacks' and rather provides a way of thinking about how a particular set of trees cannot exist without the cultural-social-politics that sustain them.

My hope is to problematise the notion that we can 'know' the forest in any complete sense. Yet, to make politically meaningful analyses of community forestry, we need a shared sense of what exactly the 'forest' is. I illustrate some 'local' understandings of forests, but these interpretations are rife with my own culturally embedded notions of a 'forest' and how it might be important to the people in Mugu. The resulting narrative is thus somewhat fractured but I suggest that it is those very fractures that allow new understandings of community forestry to emerge and for the importance of how forests are constructed by a diversity of actors to become clearer.

The cultural politics of community forestry

Community forestry in Nepal is upheld internationally as a model of resource management that has the twinned goals of providing resources for the poor and conserving forest ecosystems, in particular trees. The programmes were conceived in the early 1970s by international donors working with Nepali specialists influenced by THED (see Metz, Chapter 1 in this volume). Early community forestry pilots had limited success but nevertheless, were swiftly adopted as a national model supported by international donors who implemented programmes throughout the Middle Hills. Almost 30 years later, the programme continues to expand.[1]

To understand how forests are the centre of a set of mutually constitutive relationships with the state, communities, international donors, and user-groups, I situate Nepal's forests within discourses of environmentalism and development. In the early 1970s, fuelled by the burgeoning environmental movement and neo-Malthusian ideas of population growth, international experts expressed concern over the apparent rapid degradation of Nepal's forests (Eckholm 1975). Deforestation was believed to cause massive landslides – which often further reduce forest cover – and to increase soil erosion from Himalayan hillsides. This so-called Theory of Himalayan Environmental Degradation (THED) was quickly accepted by the international community and a number of programmes were implemented to reverse the catastrophic trend.

Community forestry was thus key to preserving Himalayan ecosystems by stopping felling and encouraging tree planting. It had financial and technical support from the World Bank, FAO and the World Wide Fund for Nature with Australian and Swiss projects pioneering plantations and later community forestry projects (Gilmour and Fisher 1991). Initially the projects were designed to encourage reforestation and to protect existing stands, but the need to provide resources for 'subsistence' agro-forestry users was incorporated into the language of the legislation from the beginning. In many ways, the programme's emphasis on assisting the poor was a long-term environmental rather than social goal. Poverty is believed to be the cause of environmental degradation and therefore reducing poverty is a key component of any conservation strategy in the Third World (Pitt 1986). Thus, both discourses of poverty and THED ensured that community forestry was from the beginning embedded in an understanding of Himalayan ecosystems as fragile, under threat from users, and in need of intervention to prevent catastrophic damage.

There is no question that there are large-scale landslides, significant run-off of top soil, and in some places, a decrease in forest cover (Ives and Messerli 1989). In this sense, community forestry was a reaction to ecological conditions and that shaped the goals of the project in significant ways. Healthy ecosystems at the time were theorised as those in dynamic equilibrium such that when a disturbance does occur, the system quickly returns to its initial state. The mountains were seen as fragile, containing rare and sensitive plants and animals that would not respond well to catastrophic change. The fragility of the Himalayas thus demanded immediate action to restore equilibrium and prevent more catastrophic change by stabilising steep slopes. One answer to this problem was community forestry; the programme went from a few pilot projects to national policy within a few years time. Within ten years, THED was discredited by geographers working in Nepal who argued that much of the degradation seen today is the result of heavy taxation burdens during the nineteenth and early twentieth centuries combined with a landscape that is inherently unstable (Mahat *et al.* 1986b). While this countertheory is widely accepted today by most Himalayan scholars, the image of massive erosion and deforestation in Nepal continues to dominate the popular imagination.

Community forestry programmes run by international donors proliferated across the more accessible areas of the Middle Hills beginning in the 1980s. Donors provided expertise, staff, and resources to user-groups, working intensively with them to form user-groups as well as plant trees so that forests could be handed over by the District Forest Offices. As the programme has developed, the social aspects have become far more significant. In part, this was due to the ways in which antecedent social relations intersected with ecological conditions in the context of specific projects (Nightingale 2003b). Many early projects were plagued with non-compliance by villagers, leading to poaching, apathy in protecting plantations from grazing animals, and a failure to properly tend seedlings. Research into why villagers were not enthusiastic revealed that in many cases project implementation had failed to adequately include local elites (Gilmour and Fisher 1991). Without their support, other villagers did not feel compelled to (or were afraid to) assist the projects and thus there was little chance of their success. These failures required project planners to re-evaluate their strategies and the result was more emphasis on community participation and inclusion of local elites.

In many ways, community forestry in Nepal has exemplified the best in participatory development. This is due in part to the ways socio-natures were produced within the programme. Most donors took an impact view – what impact did the cultural conditions have on the success of community forestry? – and yet, in practice the distinction between social and environmental aspects was less distinct. It became clear early on that changes in environmental conditions could not occur without careful attention to local social relations around forest use and protection. Thus elites were enlisted into leadership roles, and the importance of including women and lower castes was recognised (Kharel 1993). In many communities in Nepal, women and Dalit men do most of the harvesting of forest resources and yet they are often excluded from any kind of formal decision-making. But what is less clear is the extent to which the projects take the knowledge and beliefs of villagers seriously as resources for developing good management of forests. I have argued elsewhere (Nightingale 2005) that community forestry is predicated on the belief that villagers need to be taught and the programme is suffused with ideas of scientific forestry and the dissemination of expert knowledge instead of an emphasis on fostering local knowledges.

Not only is the relationship between social-cultural relations and ecological conditions more complex and mutually constitutive (as opposed to impacting each other), but the issue of knowledge is central to understanding this embeddedness. In some sense, each group of actors within community forestry interacts with a different forest due to their understanding of it. For some, the forest encompasses the management arrangements, daily resource extraction practices, the trees and the set of processes that are required to keep them intact. For others, the forest is an area delineated on a map with a set of legal obligations and permissions relevant within those boundaries, and for yet others, it is simply a collection of particular species of trees and a few associated understory species. Of course many of these understandings overlap and it is difficult to identify specific individuals who hold only one, pure version of these stereotypes. Nevertheless, it is these understandings that produce 'a forest', and the ways in which they are different in part produce the conflicts that emerge.

What is more difficult to explain, however, is what drives the whole process. Can changes occur in the politics of forests 'in isolation' from the ecological aspects? I would argue that they can. Community forestry is a good example of this. The programme was instituted nationally and has led to dramatic effects on the 'forest' in Mugu, and yet the forests in Mugu had virtually no input into the formation of that process – their ecological conditions were not the driving force behind the national development of community forestry. I feel reasonably confident saying this since the first aerial photos were taken of the area in 1978, after community forestry legislation was enacted, and very few people (Nepalis or foreigners) from outside the region had (or have) visited. It was the forests closer to Kathmandu that drove community forestry in the most direct sense. Yet community forestry has produced a series of locally important forests in Mugu that are only sustained through specific ecological, social and political processes. With this theoretical ambivalence in mind, I turn my attention to a specific user-group in order to investigate some of the local-scale contestations and struggles over knowledge and practice that occur – and produce a 'forest'.

Community forestry in Mugu

I worked in Mugu District of North-western Nepal over two time periods, 1993–94 and 1998–99, with user-groups in one valley in addition to interviewing a host of actors within the District Forest Office (DFO) and international donor agencies in Kathmandu, Surkhet and Nepalgunj. Mugu is an interesting place in terms of community forestry because it is one of the very few places where there is not significant donor input into user-groups.

I worked intensively with the first user-group to be formed in that part of the country, in 1992. The group is composed of people from three different villages and four different caste groups, roughly corresponding to the three villages. Caste distinctions historically defined the division of labour and caste also carries important social restrictions related to ritual pollution. High castes will not take food or water from lower castes, particularly the Dalits. Forty-two households of Bahuns and Thakuris live in Chaina near the district centre town of Gamgadhi. Twelve high-caste Chhetri families live in Hernikanth, a 45-minute walk up the hillside near the forest boundary. And 24 Dalit Kami or blacksmith families also live on the forest boundary, in Sangkhola, but on the opposite side of the ridge (see Figure 4.1 on p. 93). They manage Pipledi forest which covers the ridge at the top of a large, well sheltered valley with a warm, relatively dry micro-climate. The top of the valley levels off into the Rara Lake National Park, Nepal's largest freshwater lake.

Caste relations are central to how power operates within the community forestry user-group. The two highest castes, Bahuns and Thakuris, share a variety of kinship and ritual relationships. There are no Bahun priests in the village, and the Bahuns will intermarry with the Thakuris, but overall, they have worked to protect both their ritual high-caste status and also their economic dominance by controlling the better agricultural land and ensuring their sons receive sufficient education to obtain jobs in government offices. The Chhetris are also a relatively pure caste. In western Nepal, there are Matwali Chhetris, or alcohol drinkers, who are of a lower status than the Chhetris in this user-group. Oral histories explain that a variety of bonded labour arrangements predominated until relatively recently. Presently, the Bahuns and Thakuris no longer have the same control over the labour power of the others, although the Kamis still work for very low wages during times of peak agricultural work. Changes in the economy and political situation have been instrumental in breaking down caste obligations even if caste boundaries remain rigid (Nightingale 2005). These changes in historical relationships are extremely important for understanding the allocation and control over forest resources today.

The history of forest management in Mugu

Historically, land tenure and taxation in Mugu were governed by the *thekthiti* system similar to other parts of the upper Karnali. Under this system, the land was officially owned by the state, except where private ownership claims were recognised. The local *mukhiyā* or tax collector had significant powers to set the amount of tax paid by residents along with other political powers (Regmi 1988). The present *mukhiyā* has a document issued in 1911, detailing the terms under which a large tract of forest was granted to his great-grandfather to use and manage but ownership remained with the state. The *talukdār*, a government official and village tax collector,[2] regulated the cutting of trees for timber and oversaw the management of other forest resources. The document lists detailed harvesting rules, including that only dead and downed wood could be used for firewood and fencing, and that animal shelters were prohibited and slash and burn cultivation was illegal. Hunting was limited, and timber trees were to be selectively cut.

Decisions about harvesting wood were to be made by the *talukdār* in consultation with the village elders. No money could be collected for timber permits, but it was customary to give the *mukhiyā* and other village elders gifts of yogurt, beans, rice and milk. Finally, and perhaps most importantly, the document gave the *mukhiyā* the right to exclude other villages from using the forest. The oral histories all confirm that this is the system that managed forests prior to their nationalisation.

At this time, there were also private lands within Pipledi, and large portions of the forest had been cleared for cultivation with potatoes, winter wheat and millet by families from Chaina and Hernikanth. High-caste families, primarily from Chaina, built small houses and herding shacks, and moved to the forest during the summer months; today old terraces are apparent throughout the southern 20 per cent of Pipledi.

In 1918, shortly after the *talukdār* was officially given control over Pipledi, a cholera epidemic ravaged the valley and killed at least 10 per cent of the inhabitants. As a result, the lands in Pipledi were abandoned as most families did not have enough people to maintain the upper fields. Pipledi continued to be an important source of forest products and summer grazing land, but the fields quickly re-colonised with blue pine (*Pinus wallichiana*) and thick, even-aged stands of trees exist on many of these old terraces today. Subsequent land surveys reclassified these terraces as forest land, thereby stripping their owners of private property claims.

Management of forests under the *talukdār* prevailed until 1957 when the forests in Nepal were nationalised. This date is universally cited by villagers in Chaina and Hernikanth as the date when they lost control over the forest. It is likely, however, that *talukdār* management continued until somewhat later because many people who could give me details about forest management under the *talukdār* system are in their early forties. Also, according to the current *mukhiyā*, the forest was first brought under state control in 1967, and all old documents pertaining to community management of forests were declared invalid at that time. According to the *mukiya*, when the land survey was done in Mugu in 1976, all land with trees on it became property of the state, dispossessing the people of Chaina and Hernikanth from their land within Pipledi. The loss of these terraces was a significant event and one which was important for catalysing the acceptance of community forestry years later. The role of the *mukhiyā* (formerly *talukdār*) as village leader was slowly eroded during this time due to his diminishing control over forests and other village resources, and in 1981 his role was abolished and confined to the collection of taxes only.

Oral histories tell of forest conditions under the *talukdār* that were dense, wild (*jaṅgali*) and plentiful around 60–70 years ago. Significantly, one 74-year-old man tells of forests that extended approximately 215 metres lower down the valley than they do today. For most people, the forest was a frightening place because of wild animals, other people and the dense trees. Today women are still reluctant to go into the forest alone for fear of men, spirits and animals, but most people say it is not the wild (*jaṅgali*) place it once was. Many spoke of how much easier it is to collect key resources although there are important differences in the narratives across caste lines:

> *Thakuri man*: When I was 10–15 years of age the forest was very close . . . it was very easy for us to collect firewood there and we didn't need to cut green [live] trees . . . There were a lot of wild fruits such as *bhel* and *cutro* . . . There was less population [so there was] sufficient grass in the forest for the cattle to graze and a lot of fodder grass was available.

But most often, when asked to describe the forest, interviewees responded by detailing the management practices, for example: 'During those days we had to get permission from the headman. We had to get permission for foliage, fodder grass and wood. . . . [we would] take grams, lentils, curd as royalty.'

> *Thakuri man*: I remember the best as being that everyone (each village) had their own forest. But people watched the forest very carefully, there was protection. If someone asked to cut four trees they didn't let them cut five . . . If we told someone they could cut four trees in Pipledi, on a fifth tree, they didn't even touch a branch.

When interviewing people about the history of the forest, I often had to ask follow-up questions to obtain responses about the ecological aspects of the forest because people immediately began comparing the *talukdār* or national forest management regimes to community forestry. For example,

Kami woman: Our Sangkhola's forest was that one [Pipledi] from a long time ago, it was our three villages', it wasn't anyone else's . . .

Andrea: How was the forest? Was it thin or thick?

Kami woman: At that time it was thick, it has become ruined (*katum*). Now there are the saws [cross-cut saws], they have ruined the forest.

Initially I thought part of the problem was with my questions, but sometimes people would respond by talking about ecological conditions and other times even if I did ask specifically about the trees and other plants, the answers would be either very cursory or solely about management. For instance, one Thakuri man brushed aside my question about historical forestry and discussed the present (community forestry) management.

Andrea: In your recollection, what did you like least about the forest?

Thakuri man: [He answers in a different vein] What I like is that we have our own forest, with its clear boundaries. What I do not like is if we do not have many rights to protect the forest. There isn't anything bad to say. But, what we didn't like is that after the Rastriya *ban* [National forest] was formed, no one protected the forest and there was a lot of loss.

This response was part of a series of questions in which I tried to elicit information about historical ecological conditions but he continued to focus his responses on the present. Similarly, many people segued from the *talukdār* system, to the national forest, to community forestry in one breath:

> *Chhetri woman*: Before, the forest was really big, there were big trees and then all of them were cut. If we asked why they were doing that, they would say, 'It's the government's forest.' At that time we could not say anything, 'Do not bring baby goats here,' we could not say that . . . Before, people were careless, they would cut seedlings, cut branches . . . we thought, 'Oh, we also have sons, they need [the forest], are we going to do this?' And we saved the forest, for that we made our own forest.

Thus, for the villagers, the forest is not simply a collection of trees or a place to which they can go for resources, it is the confluence of the trees, resources, management authority, and the harvesting practices of themselves and others. It is meaningless to speak of the forest without referring to these other aspects as indeed, the very definition of 'forest' in the local vernacular carries an implication of usufruct rights, uncultivated land and a place where fodder, grazing, fuel and if the trees are big enough, timber are available. In many ways, theirs is a relational understanding of the forest such that any description of it must include at least social-political as well as ecological dimensions and their sense of whether resources are plentiful or not is closely linked to the extent to which they feel in control of access rights and the levels of harvesting. Thus one Kami woman said, referring to the national forest,

> Before there were enough pine needles and other things. Wherever you went, there was forest. We could not eat all the peaches we got, we had to throw some away . . . now what to do? There is not a forest like that any more. They [people from Chaina] dominate us (*hepchan*), they do not give us permits, we cannot even talk to them. We always have to do their work.

I spoke to her soon after an incident where the Chaina women caught the Sangkhola women harvesting leaf litter out of turn, and thus her sense of being dominated was particularly acute. But other men and women from Sangkhola also lamented the loss of the national forest system because they had found it easier to transgress the rules and avoid paying for key resources. For them, resources were 'easier' under the national forest because they did not have to go through the higher caste people to get them even though they can see resources regenerating now.

Regardless of their perspective, all respondents wove together narratives of ecology, access to resources and comparisons between the old system and the new in their oral histories. Such a forest is difficult to map because these systems of control do not lend themselves well to spatial visualisations. The people who did offer to draw maps of the old forest were careful to point out which areas had been subject to the most change in terms of management, which was often implicitly (but rarely explicitly) linked to changes in forest cover.

The origins of community forestry

Another aspect of the oral histories highlights the ways that the 'forest' can vary depending on people's positionality. The stories that people tell about the community forest orgins are closely linked to their current interests. Thus in some sense they are told in relation to different forests. These interests are bound up in ecological conditions and people's livelihood strategies. I examine how different knowledges are employed for strategic interests, making them mutually constitutive of social relations of power within the group and between the group and the District Forest Officers.

The prime motivation for starting the community forest was the degraded ecological conditions in Pipledi. User-group members talked passionately about how Pipledi was overused after the nationalisation of the forests, primarily because they could not exclude outsiders and regulate the use of resources. Yet different groups of respondents, generally falling along gender, village and caste lines, emphasised their roles in initiating the process differently. I was told several contrasting narratives, none of which could be conclusively corroborated. This ambiguity is representative of the struggles over social relations. By claiming responsibility for initiating the community forest, people assert their own interests and rights to control and manage the forest. They make these claims by demonstrating knowledge of ecological degradation, scientific forestry and development, and in the case of the Dalits, by claiming ignorance of these, thereby allowing them to behave as they see fit.

According to the Thakuri women, the community forest was formed at the initiative of the community – specifically Chaina.

> Long before, when our forest had finished [was degraded], we had the idea [to protect the forest]. Then we heard on the radio about other places like Rolpa [a district which has community forestry]. Three rangers and Chaina, Hernikanth and Sangkhola got together and made the decision. Manohari [a political leader] helped a lot. At that time the committee was 13 people including four women. When they were in community forestry meetings, the men would applaud the things the women would say. [The woman speaking clapped her hands, demonstrating.] They thought our ideas were good. The men were educated (*paṛe lekheko*), we were not, so we needed both men and women for the community forest.

The Chaina women emphasise their role in the formation of the user-group. They blocked the men's proposal for firewood permits, arguing it would be an undue burden given they collect firewood daily. Here they are making claims to expertise in managing the extraction of forest products, but they make it clear that they could not handle the administrative aspects of community forestry. Legally, the group must develop a written management plan and keep accounting records to take control of daily management. The women are not literate whereas the men are, so both these kinds of knowledges were seen as important for producing the community forest.

The Thakuri men, in contrast, say they heard of the programme from the forest rangers and a political leader from Chaina persuaded the others to form the group. 'The DFO . . . told us about CF and helped to start the process.' Another man said,

> In the beginning we did not know anything . . . One day he [DFO] asked all of us literate people (*janne manche*) if we want to take our forest. We did not understand, but he talked with our older brother about this. After that our older brother said, 'OK, we do not understand, but we are ready to try it.'

According to these high-caste men's accounts, the District Forest Officer was the person who originally promoted community forestry and the villagers were reluctant to trust him. It was the wisdom and foresight of one of their village leaders that allowed them to accept the programme.

In these narratives, there are two important threads. One is the question of who learned about community forestry and who was willing to try the programme. The women gave the men credit for convincing them to try it but claim independent knowledge of it (from the radio). The other is that of who understood what community forestry might mean for the community and how to implement it. The women insist they made very important contributions to the development of the management plan even without being able to physically write it. Theirs is a practical knowledge, embedded in their daily use of the forest. The men, in contrast, are careful to emphasise that the DFO was the expert and that the village leader was the key person with the foresight to try the programme but none of them really 'understood' what it meant. In many ways they portray themselves as lacking knowledge both of the programme and of managing the forest itself (Nightingale 2005).

These claims about knowledge and a lack of it are integral to how people seek to position themselves in relation to various kinds of authorities. The men have a vested interest in cultivating a good relationship with the DFO staff and to do so, they need to be appropriately deferential and make it clear that they see themselves as inferior. But importantly, they also do not need to prove (to me or anyone else 'outside') that they are the rightful managers of the forest; their literacy skills, high-caste status and gender ensure that claim under local cultural norms. The women in contrast generally have few direct dealings with DFO staff and have a greater need to emphasise their knowledge and thus claim to managing the forest in part because they have become largely marginalised from the committee management process (Nightingale 2006). Many women stated that women's influence on the committee has deteriorated over time as the leaders have become more entrenched and disputes between political parties have taken precedence over other alliances and issues within the group.

The Kami men and women essentially corroborate the Chaina men's narratives, but they simply credit Manohari with starting the process. It was their trust in and deference to Manohari's leadership that allowed the programme to begin. I think their reluctance to give me details about the process not only reflects their marginalisation during the entire formative process, but also their lack of trust in me. I found it difficult to elucidate a distinctive Kami perspective on many community forestry issues which I believe reflects the difficulties I had in developing relationships with them and this should not be taken as an indication of their lack of knowledge. It is also important to recognise that a political leader of Manohari's standing historically would have controlled the forest, and the Kamis believed they could not contest his authority over it; therefore it is not surprising they did not seek to do so to me.

The Chhetris from Hernikanth (men and women), however, do tell a different narrative. They emphasise the role of a man from their village who saw publications in the DFO office and approached the DFO staff about forming a group. In an interview one man said, 'In the beginning, I was at the DFO office and I read about community forestry in a magazine that was on the accountant's desk. I picked it up and asked the accountant what it was about. The accountant explained it.' Another said, 'Actually our village is the only one to start this [community forest]. It was the secretary and the teacher's father, then after that Chaina found out about this and they made Manohari [from Chaina] the chairman. We ourselves started this community forest in the beginning.' It was only later, according to this account, that Chaina's community forest and Hernikanth's were joined. In the first few months, Hernikanth and Sangkhola were said to have had a community forest that was separate from Chaina's. No one from Sangkhola or Chaina professed any knowledge of this initial arrangement.

I believe this claim is related to current struggles within the group and the desire of people from Hernikanth to take over managing part of the forest independently. People in Hernikanth rely heavily on animal husbandry as a source of subsistence and cash and feel the forest is not managed sufficiently for grazing. The inventory work I did corroborates this; the current management practices are largely targeted at producing good timber rather than grazing vegetation.

The women from Hernikanth also discussed the role of their village in forming the community forest. Like the higher-caste women from Chaina, they pointed out the importance of their own knowledge, but they framed it somewhat differently than the Chaina women, acknowledging more fully their own initial reluctance:

> We were afraid. Women scolded their husbands and asked them why they were forming the community forest and the rules [in the way they were]. The men gave their wives a lot of trouble. The men didn't know how to manage the firewood, they wanted to have permits for that too. The first time that they wanted to do this, the Chaina, Hernikanth, Sangkhola women who didn't understand scolded their husbands. Not all of us [laughs], but it took some time for some women to understand. Now we all understand.

This observation is difficult to interpret but I believe it means that the men did not have a good grasp of appropriate harvesting rules, and that the women did not appreciate the *value* of community forestry which produced conflict between husbands and wives. The Hernikanth women present a rather self-deprecating image of themselves. They equate understanding with development and recognising the value of a programme like community forestry (see also Pigg 1996). Yet they simultaneously highlight the importance of their knowledge of harvesting practices for key resources, especially firewood and leaf litter. In sum, they suggest they gained an understanding of forest protection through the community forestry process, but also insist on the relevance of their prior knowledge for managing the resources they work with every day.

Thus people from Hernikanth, by insisting on their knowledge and activities in initiating community forestry, assert their claims to Pipledi and its management. These claims are very important. It emerged later that many people in Hernikanth would like to divide Pipledi and manage a smaller piece themselves. Several men mentioned that they would like to split from the larger user-group, because the people of Hernikanth were more cohesive in their ideas and could do a better job managing a smaller forest. The women mentioned that there is not enough grazing land and they are concerned that the re-growth of blue pines in Pipledi will not help create more fodder (they are 'correct' on this point from a forestry perspective). One way to promote their agenda of splitting the forest was to emphasise to me their role in the process, and establish their temporal precedence in forming the community forest.

Questions of expertise and control were not confined to the formation of the group. Different kinds of knowledge of forests and questions of expertise were central to daily conflicts and decision-making forums when I did my fieldwork. The high-caste men in the user-group have a vested interest in cultivating a strong relationship with the DFO staff. Historically such relationships have proved useful to them and many see connections with government employees as a potential source of jobs, support in local disputes and other intangible benefits that may accrue later. One way in which they seek to curry favour is by demonstrating that they are educated and 'developed'. Most often they do this by invoking understandings of scientific forestry and deferring to expertise when arguing for particular management strategies. As one Bahun man said, 'Until 2034 [1977] all the forests were Rastriya *ban*. Then scientists, specialists, and scholars came from different countries and thought that the forest was for the people. And it is better to give the right to the people to conserve their own forest. If you put it in their hands it *is* better.' Thus, high-caste men often negated the importance of their own knowledge and utilised discourses of teaching and subservience to indicate their deference to the learning of the DFO staff and international experts who advise them. Because of these hierarchies, by appealing to their trust and understanding of the DFO's forestry practices, high-caste men are able to assert their superior right to control the user-group.

Ecological elements of forest dynamics

The narratives above are one way to understand the history of community forestry in Mugu. I now want to provide a different analysis using aerial photo interpretation and a vegetation inventory. The forest that emerges is very much an 'ecological' forest and most issues of control, access and politics are absent. Yet, it is the following kind of analysis that is used to evaluate whether Nepal's forests are 'improving' and to develop policy for future management. I want to contrast these different 'forests' to highlight the importance of integrating different kinds of analyses and to show the cultural constructions of knowledge.

Forests in this part of Mugu are clearly in transition due in part to the National Park and community forestry programmes, both of which have changed forest use practices. Areas within the park have regenerated and areas outside show both clearing and regeneration. Forest dynamics are also driven by a variety of ecological factors, but it is difficult to assess their relative influence without detailed data to calculate the growth and recruitment rates of forest species. Historical ecological work is challenging in Mugu as virtually no detailed vegetation inventories have been done for the Karnali Zone with the exception of the present study. The oldest available aerial photos were taken in 1978, and another set in 1996. They provide some data on forest cover change, but their scale is so small, 1:50,000, that identification of crown cover species is impossible and the topographical distortion is so extreme that calculations of density are outside the scope of this study.

The aerial photo interpretation was done by mapping areas of different vegetation types onto a base topographical map. I mapped areas of forest (greater than 50 per cent canopy cover), partial forest (less than 50 per cent canopy cover), and open areas. I ground-truthed the photos to be sure that what I categorised on the photos was present on the ground. I was in the field in 1999 using 1996 photos, but forests do not change that quickly unless they are cleared and, therefore in general, the photos and the landscape corresponded. I plotted these areas onto a base map, compensating for the distortion in the photos. The base map and the two layers were then scanned and rectified in ArcInfo by Dr Mark Lindberg, the University of Minnesota cartography lab director, allowing calculations of area and the cartographic presentation of the data.

The map itself is the 1964 Indian Army survey map and is not very accurate, particularly the contour lines. The new topographical maps for this area were not available in 2001 when this work was completed, making the aerial photo work difficult as areas could not accurately be placed in the vertical plane. But given the amount of distortion in the photos, absolute measurements, even after corrections, still contain a wide margin of error. These problems make a calculation of absolute area possible but virtually meaningless and therefore I have reported my results as changes in relative area. I have done my best to standardise the area I calculated for the polygons so that their percentages are comparable. The total area I interpreted encompasses the entire valley and surrounding hillsides as aerial photo work is generally done on a much larger scale than that of Pipledi alone. Therefore it is total change in different vegetation types that can be measured rather than absolute area. The results of this work are presented in Figure 4.1 and Table 4.1.

The most improvements would be expected within Rara National Park as those lands have been restricted since 1978. The photos do indeed show that

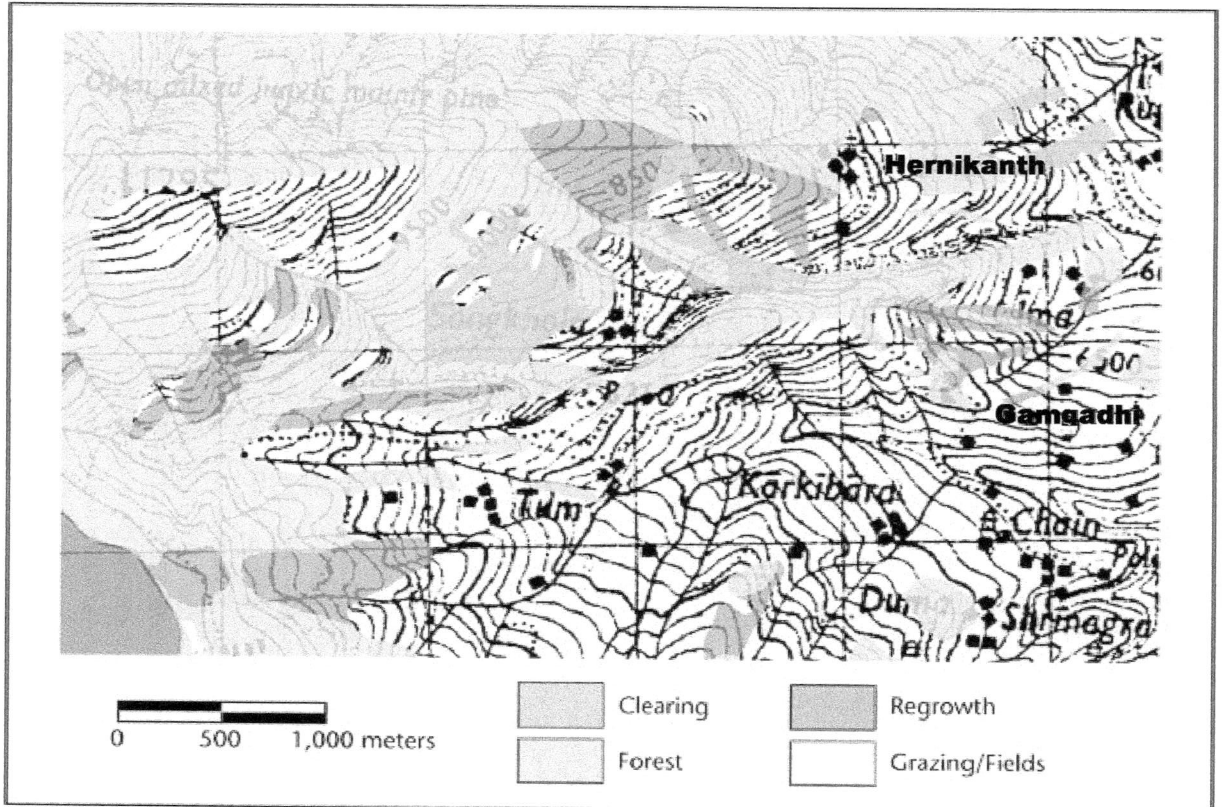

Figure 4.1 Forest cover change in Pipledi and surroundings, 1978–96.

Table 4.1 Change of forest cover types near Gamgadhi (%)

Cover type	*1978*	*1996*	*% change*
	% of total area	*% of total area*	
Forest (+ thinned areas)	28.1	33.8	5.8
Forest (– thinned areas)	28.1	32.7	4.7
Patchy forest (– thinned areas)	6.2	2.3	–4.0
Patchy forest (+ thinned areas)	6.2	3.4	–2.9

most of the open areas on the eastern side of the lake have filled in with thick forest. On the ground, one can see even-aged stands of blue pine growing in these areas. The north-eastern side of the lake also shows significant regeneration, especially adjacent to Pipledi forest but within the Park boundary.

Closer to the three villages, there is a decline in forest cover in some areas and improvement in others. The area below the lower boundary of Pipledi had been forested in 1978 but in 1996 had been cleared (see Figure 4.1). Similarly, the old terraces within the middle of Pipledi abandoned during the cholera epidemics have recently been cleared. When the community forest was established, many families were able to reassert their claims to those terraces and they were reclassified as private. The upper elevations of the forest, however, appear to have become thicker, and areas that were once largely cleared, now show relatively closed forest canopy. Interpreting density on the photos is extremely difficult and unless the change is dramatic, it can be hard to state conclusively that the canopy is thicker. Yet, in comparison to other areas that do not seem to have changed much at all, some areas of Pipledi do seem denser.

Thus in contrast to the generalised premise of THED that forest cover is diminishing, overall, the picture is mixed but shows some improvement. Forest cover has increased by 5–6 per cent; patchy forest has decreased by 3–4 per cent; and open agricultural or grazing areas have decreased by less than 2 per cent (see Table 4.1). The exact percentage of change is difficult to state because of the problems with interpreting density from the photos. I have calculated change based on treating the apparently thinning areas in two ways. In one calculation, I treat it as 'forest' and in another as 'patchy forest' thus the maximum amount of forest increase is 6 per cent while the minimum is 5 per cent. Also, the main forest polygon in Pipledi is contiguous with the Park and therefore some of the forest change observed may be attributed to regeneration in the Park.

Despite these difficulties in interpretation, there are areas close to Chaina and Hernikanth that have improved noticeably. An area on the hillside above Gamgadhi, about 20 minutes walk from Chaina has begun to regenerate. It is technically outside of Pipledi (it was included in the aerial photo interpretation), but Chaina and Hernikanth residents manage it as a community forest and are able to exclude outsiders. Similarly, the forest above Hernikanth that had been very thin forest with large open areas now has largely grown in, to the point where in many places on the photos, it looks no different from other parts of Pipledi.

Thus the aerial photos show small improvements in forest conditions, but overall very little change. In this sense they contradict the oral histories which insist that forest conditions have improved dramatically with community forestry. These results only make sense if one considers that the aerial photos show a different forest from the one the villagers know. As argued above, their forest cannot easily be mapped or captured on film because it includes a host of political, social and material practices that are not necessarily visual. Thus the aerial photos give an impoverished description of 'the forest' that can only be enhanced by the use of oral histories and local knowledge to supplement them (Nightingale 2003a).

In addition to aerial photo interpretation, I also carried out a quantitative vegetation inventory of Pipledi. The aim was to provide baseline information on current vegetation in the forest, including species composition, stand structure (i.e. the height layers in the forest), stand age, density and frequency of dominant species, and their spatial variations. This work cannot give estimates of growth and regeneration rates because it is the first of its kind for this part of Nepal. At least one other inventory needs to be done, in five to ten year's time, before rates of change can be calculated. An inventory of a blue pine (*Pinus wallichiana*) dominated forest was done in Kumaun in India in 1986 (Bargali 1986, Singh *et al.* 1994) and those data can be compared to the present study to give an indication of whether the forest falls within the range of values found elsewhere, although micro-climate conditions are likely to be significantly different, making such comparisons of dubious value.

The vegetation inventory measures a number of qualities of the forest, particularly of trees. Most of these measurements were developed for the commercial timber industry, thus when analyzing a 'subsistence' forest, it is difficult to know which measurements to use and which have much value for understanding a forest with this kind of use. The inventory was designed to be a semi-random, plot-based survey of the forest, gathering both quantitative and qualitative information on the vegetation (Avery and Burkhart 1994). The measurements from Pipledi that could be compared to the Kumaun study were primarily basal area and density. Mean basal area is one of the key measurements for commercial forestry. This measures the average cross-sectional area of tree stems at breast height (dbh) and can provide a measure of the total board feet of timber in the stand. While villagers do have a need for timber, it is also a useful measurement for comparison with other stands and to understand aspects of tree growth when other inventories are completed in the future. In Pipledi, the mean basal area per tree is 0.07 metres squared per tree compared to 0.05–0.08 found in the Kumaon study, indicating that the areas of the tree stems are within the same range. The mean basal area per hectare, however, is outside the range at 12.7 metres squared per hectare compared to 27–45 in Kumaun (see Table 4.2). Similarly the density of trees in Pipledi is 169, well below the range found in Kumaun, indicating there are fewer trees per hectare in Mugu. The reasons for these discrepancies cannot be determined without further study, but are most likely due to the degree of human disturbance and precipitation. The stands in Kumaun are described as 'old growth, apparently natural communities that are self-maintaining' (Singh *et al.* 1994, p. 405), whereas Pipledi is heavily harvested. Precipitation in Kumaun ranges between 1,000 and 3,000 mm per year, compared to the 700 mm per year in Mugu, and increased moisture would cause increased growth rates in trees.

The vegetation inventory, in the absence of a follow-up study, thus gives a rather thin picture of the forest. Essentially the average cross-sectional area of each tree is roughly similar to that of the Kumaun forest, but the total number of trees in a given area is lower, making the total basal area per hectare significantly lower.

The inventory does give some idea of the diversity of trees and other plants found in the forest, however. Indeed, I was surprised by the number of tree species (10) as the initial impression is one of low species diversity primarily composed of oaks, rhododendrons and blue pines, yet in places it is dominated by other species. A forester looking at this data would see a rather thin forest relatively typical of a dry, alpine micro-climate, without huge potential for commercial timber production. The blue pines do regenerate quickly after a disturbance and primarily grow straight and tall. They make good timber, but are less useful (yet widely used) for firewood. There is a relative scarcity of oak seedlings, which is cause for concern because the villagers rely on the oaks for fodder.

Overall, this kind of knowledge of the forest is skewed towards tree species and specifically information on mature trees and how much mass they contain (i.e. basal area, height, crown size). It does not reveal much about harvesting pressure and it certainly does not give any information on who is controlling the forest. Some information about management strategies can be obtained, however. For example, there are fire scars on the outside and deep within the cores of many of the old trees, indicating that large fires have been frequent. The relatively even-aged stands and large number of seedlings and

Table 4.2 Comparison of basal area and density of Pipledi to a forest in Kumaun

Study	*Altitude m*	*Mean basal area m2/tree*	*Total basal area m2/Ha*	*Density #trees/ha*	*Species richness (total # tree species)*
Bargali 1986	2225–2585	0.05–0.08	27–45	540–690	n/a
Pipledi	2475–3275	0.07	12.67	169	10

saplings of blue pine similarly are indicative of the recent fire-suppression practices of the user-group. There is more seedling survival due to the lack of fire. Yet making these kinds of statements about management is predicated on having some information about current practices and not easily read from the forest data alone.

Perhaps most importantly, both the aerial photo data and the vegetation inventory produce a forest that is clearly bounded in space and time. They help to give the impression that the forest is a 'place' and while a changing place, the boundaries of it are not, at least in terms of what is called 'Pipledi'. One can examine changes in forest density and discuss areas that have been cleared but this is done in reference to the place itself, as if there is some kind of blank geographical space upon which trees and other species can be mapped. This is in stark contrast to the villagers' accounts, which merge the place with the politics; boundaries and forest qualities change with the speaker and the time period under discussion.

Conclusion

This chapter has traced the history of community forestry in reference to a particular place, pulling out different stories of the forest and its management within that history. Questions of knowledge are paramount in debates about forestry, particularly as initial proclamations about the crisis for Nepal's forests have been slowly debunked by both politics and science. Studies on forestry revealed that many places did have forest management regimes in place, whether or not state-sponsored oversight was effective (Gilmour 1990) and 30 years on, Nepal continues to have a reasonable amount of forest cover, although estimates of its change have been contested (Nepal 1999; Zurick and Karan 1999). In addition, geomorphologists reassessed the soil erosion and landslide crisis and concluded that in an area of significant plate uplift like Nepal, unstable slopes and corresponding soil movement are 'normal' (Ives and Messerli 1989; Thompson and Warburton 1985). These studies have contributed to an overall change within many of the physical sciences which have begun to theorise ecosystems as dynamic and characterised by disturbance, rather than stability. These theories see human activities as a more natural part of ecosystem change and indeed, as very important agents of change (Botkin 1990). Yet what ecological science seems to fall short of embracing is that this kind of understanding requires that we reject the idea of human 'impacts' on ecosystems and rather seek to explore the interrelationship between politics, culture and ecology (Braun and Castree 1998; Cronon 1996). Some critics of THED have made similar arguments (Thompson and Warburton 1985), but most of this scholarship continues to work within impact or feedback systems frameworks that cannot capture these interrelationships adequately.

The narrative I have traced here takes these latter insights seriously and highlights how *different* forests are produced by the ways people understand community forestry in Mugu. In my PhD I sought to integrate the oral histories, aerial photos and vegetation inventory into a seamless whole. I used them in relation to each other in a dialectical fashion, assuming that each was embedded within the other and thus wove them together throughout. This allowed me to highlight the importance of control and access to the forest for villagers and to explain how forest conditions had in fact improved dramatically when the aerial photos seemed to negate that (Nightingale 2003a). It also demonstrates some of the successes and failings of community forestry as disputes within the user-group are quite literally played out on the ground, causing forest fires, extinguishing them, or resulting in the cutting of green wood for firewood.

The history I have traced here, however, begins to raise other questions about whether assuming an a priori existence of 'the forest' is justified theoretically. A study of community forestry would, of course, expect to occur in reference to a forest at some point, even if only implicitly through an investigation of user-groups. The forest also provides a neat conceptual anchor against which changing social relations, ecological dynamics, the politics of management and daily harvesting practices can be mapped both cartographically and representationally. It is the place to ground the study, quite literally.

Yet as I have demonstrated above, the place itself cannot be taken for granted. Its boundaries, qualities and meaning shift with each method used to explore it and from different perspectives, requiring one to question the existence of *a* forest. What then, is the forest? Is it a collection of tree species of importance because of its resources? Or are the resources, politics of access and control, management practices, scientific assessments and daily conflicts all required to produce and maintain the forest? It is the emergence of the culturally constructed category(s) and place(s) called the forest that can only be explained by exploring the interrelationships between knowledge, practice, politics and ecological change.

The villagers themselves, while clearly defining a place called the forest, refuse to separate the ecological conditions from the management practices and politics of control. The higher castes (Bahuns, Thakuris and Chhetris) see the current situation as substantially better than the national forest, not because poaching has been eliminated (it has not) and not because there are substantially more resources, but because now they have control over them. They can curtail poaching or demand compensation when offenders are caught, and they have a far greater sense of security knowing that they can decide how and when to use their resources. The Dalits (Kamis) are not convinced the present situation is better, however. They felt more in control and less dominated by the higher castes when it was only the DFO guard they had to worry about rather than all the other members of the user-group. In all these instances, the forest is a different forest now and they cannot talk about how things have changed without referring to both the ecological and social-political aspects of their 'new' forest.

In this chapter I have deliberately separated out the social–political and ecological accounts to highlight the different knowledges that are brought to bear on one tract of land. The 'scientific data' (aerial photos and inventory) each provide one account of the forest. Within this work, there are not multiple narratives to explore and reconcile, as they each have only one story to tell. The oral histories, in contrast, can be triangulated with the other two data sets, but only incompletely (Nightingale 2003a). Different groups of people tell contrasting stories and thus how it is that the forest can be much better and worse at the same time emerges clearly. These contradictions highlight the complex relationship between knowledge, daily practice and the processes of development evident in community forestry.

Community forestry has indeed provided resources for the poorest of the poor and engendered a strong conservation ethic within the user-group in Mugu. Development processes have produced forests both within Mugu and nationally and drawn many people into development agendas in contradictory ways. Knowledge of forests and how this knowledge is told, mapped and contested are central to the production of 'developed' and 'undeveloped' people within Mugu and thus both ecosystems and social politics are at stake in community forestry. This account, then, goes some way towards refuting the validity of THED not only because the knowledge used to understand ecological change is not consistent, but also because it is clear that some losses and some gains have been made within the socio-natural complex of forestry in Nepal.

Notes

1 Several donor-funded projects have closed since 2003 but the government continues to hand over forests to user-groups.

2 *Mukhiyā* and *talukdār* are roughly synonymous although in many circumstances, the *mukhiyā* had authority over several *talukdārs*.

5 Where God's children live

Symbolizing forests in Nepal

Jana Fortier

If people in Nepal think about, and symbolize, their forests in culturally distinctive ways, then precisely what are these distinctions? And why are different visions of nature important? In this chapter, I conclude that there are fascinating and multiple meanings assigned to forests and forest spaces by the foragers and farmers living in the Himalayan foothills of western Nepal. Specifically, the forest-dwelling Raute (pronounced "Rao-tey") conceive of forested places as a home, a domestic environs, in which live the children of God: the beings living in the forests. A society of hunter–gatherers, the Raute specialize in hunting langur and macaque monkeys while foraging for about one hundred varieties of edible plants through each gathering season. These food collectors live in the Karnali River watershed of Nepal and along the Maha Kali River bordering India. Their Tibetic language is closely related to that of other local part-time hunter–gatherers, the Ban Raji who specialize in porcupine hunting, the Raji who favor fishing, and the Chepang who are renowned for their bat hunting skills. Rautes also share many cultural practices with other regional hunter–gatherers, such as the Birhor and Hill Kharia of northern India (Fortier 2009a, 2009b). According to census figures, the population of the Raute groups includes about 700 Rautes, 2,500 Ban Rajis, and 2,500 Rajis. In addition, there are about 15,000 people known as "Raut" who have assimilated into Nepalese caste society in the last few generations but who are descendants of forest foragers.

Residing in a forested environment, Rautes maintain a profound respect for forests and the resident wild animals. Rautes maintain that all the forest beings live within a home peopled by yams, monkeys, birds, small trees, stones, and other sentient beings. Rautes maintain a deep sense of their relationship to these social others since together they are the children of Berh, the Sun deity, who has other incarnations such as Gwahing and Damu. In this forested world, Rautes maintain a familial sense of kinship with other non-human beings since together they represent what they describe as "the children of Berh."

Necessarily, Rautes share the forests through which they migrate with many other people. For local farmers, the sylvan environments evoke fundamentally different ideas, based upon Hindu interpretations of the natural environment. Hindu-based ideologies of forests as natural/supernatural spaces are interesting since they are so different from Raute worldviews, and they form a background for reflection of the forest foraging Rautes' beliefs. This chapter will focus on the Raute's beliefs about themselves, the animals of the forest, and of forest spaces, but for further reading concerning ideologies of forests, see Arnold (1996), Dove (1994), Jeffery (1998), Parajuli (2001), Bhatt (2003), Cederlöf and Sivaramakrishnan (2006), and Skaria (1999).

The right to define a given locale is not arbitrary; different groups have vested interests and political power over others and thus have an interest in asserting the correctness of their vision of the landscape (Humphrey 1995). In rural Nepal, particular landscapes, such as community forests, come to be known by Nepali villagers through the subjectivity of possession. Community forests are figured as *hamro ban*, "our forest," and community user groups aggressively protect their rights to particular parcels of forested land even though the Government of Nepal legally owns most forests. Rautes, on the other hand, have no vested interest in owning or protecting a particular forest. Their concern, rather, is in the utilization of its products, in the image of the forest as a nurturing being, and in maintaining a sense of gratitude toward the forest as a parental figure.

Semantic differences concerning forests have fundamentally important political ramifications because both foragers and farmers must negotiate multiple and contrastive claims to ownership and authenticity of forested domains. Negotiating not simply the material dimensions of forests (resources, land mass, biomass quality) but especially sets of meanings, forests become objects of political contention. Are forests to be used as camps, retreats, preserves, wilderness areas, sustainable growth areas, or ethnoecological refuges? In the twenty-first century, forests in Asia are sites of contested ideologies that face the political pressures of logging companies, government development schemes, and urban sprawl (Guha 1990; Murashko 2000; Peet and Watts 2004). For people everywhere, the forest question looms large as we contemplate the role of forests in contemporary society and environmental policy-making.

For forest foragers whose subsistence depends on hunting and gathering, forest protection is tantamount to cultural survival. Foragers de facto live in the forests and view forested places as their domestic space. Foragers give forest places toponyms that highlight notable natural and utilitarian features. For example, Raute foragers name forests according to where berries and yams are found, where bears live, or as a place that is the shape of a particular leaf. In their Tibetic language, they can also give forests higher-order names to reference a midland temperate forest (*manāng*), found in Jajarkot District or a hotter sub-tropical forest (*damār*), such as found in Dang District. Rautes forage through many environments and districts of western Nepal, including Jajarkot District where my fieldwork with both Rautes and Hindu farming households took place over the course of two years in the 1990s. Yet local farmers use none of the Raute names, instead naming the nearby forests according to their own toponymic systems.

For millions of forest-based farmers living in the Himalayan hills, gathering wild tubers, berries, and greens is important for supplementing their subsistence diet. Manandhar and Manandhar (2002) record an astonishing range of forest plants used by farmers in Nepal, including 1002 medicinal plants, 651 food plants, and 696 plants used for thatch, fodder, furniture, religious paraphernalia, etc. But while both foragers and farmers share their vision of forests as places to satisfy subsistence, the states of Nepal and India view forests in more proprietary terms. In Nepal, forests historically have been viewed as opportunities for taxation of its saleable products. Farmers in Jajarkot District, Nepal, where I conducted interviews with farming householders at various times between 1986 and 1993, report that tax collectors often took all edible forest products that exceeded householders' subsistence. The tax collectors even taxed farmers by taking a portion of the wild yams they collected. More recently, however, government officials such as District Forestry Officers have focused on working with community user-groups to reforest and protect endangered forest areas.

How can the changing meanings of forested places be negotiated among various interested parties such as foraging farmers, hunter–gatherers, and government officials? Especially between dominant polities and subalterns whose subsistence depends on hunting and gathering in the forests, views of society and nature are often diametrically opposed. Raute foragers, for example, view wild yams as food but also as their own ancestors. Farmers, on the other hand, see yams as a tasty addition to the evening meal and probably will not adopt a view of yams as sentient beings. Some scholars doubt that a post-humanist approach in which the distinction between society and nature can be dissolved is possible (Howell 1996). The anthropologist Signe Howell is correct in pointing out that most people from agricultural societies will scorn belief in the environment as a sentient being and are unlikely to believe in the sentience of yams, stones, and other natural objects. Other scholars, however, find people *must* develop a more nuanced understanding of their role in the material world (Hirsch and O'Hanlon 1995). Such a perspective need not incorporate beliefs in a sentient ecology *per se*, yet at the same time a more complex perspective acknowledges the ecological role of humans within the physical environment. Ultimately, whether environmental policy-makers address such ontological issues is important because perspectives of humans in the "natural" world shape policies that affect forest quality and coverage. Ultimately it becomes necessary for us to revisit cultural attitudes towards forests and to rethink how beliefs about nature affect forest initiatives such as community forestry projects.

Not only is the perceived relationship of humans with nature at issue, but also the historical problem of the social construction of nature itself continues to be a topic of contention (Arnold and Guha 1995; Descola and Pálsson 1996; Escobar 1996; Ellen and Fukui 1996; Gell 1995; Hirsch and

O'Hanlon 1995; Ingold 1986, Viveiros de Castro 1996; Williams 1980). All of the debates concerning society and nature cannot be reviewed here, but I do wish to comment on several salient points concerning the social construction of nature as it plays out in contemporary South Asia. First, even though paradigms of nature and society are increasingly ruptured by modern political events and disasters (Blaikie and Brookfield 1987), notions of nature as wild and outside the bounds of the 'social' continue to form the basis of modern political policy-making in Nepal and India.

In particular, the notion of things as *saṃskrtik*, or cultural, acts as a political tool to enforce a set of hegemonic ideas about forestry use that are employed by the dominant and technologically complex societies of South Asia. Villages, fields, orchards, and roads are cultural places in which villagers invest time, labor, and money to build and maintain (cf. Russell, Chapter 3 in this volume). Using a policy of social development (*samāj sudhār*) based on the privilege of *saṃskrtik* social rules, dominant polities control technologically simple societies of forest dwelling peoples who are contained within their national borders. Local places as well as populations are thought of as *prakriti*; a forest is conceived as *prakriti*, or a local dialect is called *prakrit*, meaning the 'natural' language of the local people. Things and people deemed *prakriti (prakrtik)*, natural, will thus be subject to the hegemonies of their more *saṃskritik* neighbors as long as there is a political need for it. The basis for hegemony of culture over nature stems ultimately from a philosophical balance found within the *Samkhyā* (*Sankhya*) notions of *prakriti and puruṣa*. In this religious discourse, *prakriti* elements act as a form of active power while *puruṣa* lies dormant as a potential, but inert, authority. Broadly, *puruṣa* is activated by *prakriti* and the two metaphorically embrace as equals. In sociological discourse, *prakriti* and *puruṣa* emerge in many significations, with *prakriti* signified as female, powerful, natural, active, uncontrolled, wild, undirected. As Wendy Doniger has pointed out, *puruṣa* signifies male authority, a kind of seed and carrier of social hierarchy (Doniger 1980: 118). By extension, *purusa* signifies control, cultural authority, *saṃskriti*, civilized behavior. During everyday conversations by Nepali villagers, these symbolic distinctions are not consciously brought out in folk-philosophies of the forest. But these principles are applied in some social situations (weddings, sacred thread ceremonies), and variations of these significations ripple away from deeply held Hindu philosophical models of the universe.

In South Asia, in particular, there remains a swirling set of polysemic messages around the ideology of things natural or *prakriti*. The concept of *prakriti* contains an emergent quality that is contingent on local circumstances and individual experiences, rendering it something pure, local, natural, uncultivated but also fraught with a negative valence of the *abikāsi*, an undeveloped state of being. For example, an elderly farmer told me of a legal battle that his family won, saying with a certain disappointment: "After we won, the land was full of trees [smacks his lips disapprovingly]. The reason is because the land was never plowed."

This sense of the natural as backward or in need of improvement can also shift metonymically onto people who reside in forests. From the perspective of the people living in political centers such as Kathmandu, rural people who live in or near forests are framed as "backward." In western Nepal, years of hegemonic subjugation to urban centers in Kathmandu and India have resulted in the sense of rural householders as marginalized and backward, and as people needing upliftment and improvement (Pigg 1992). This sense of superiority of urban over rural has been found throughout the world and may be a global practice (Shields 1991), yet its cultural contours are specific in South Asia to the negative ideals that are associated with *prakritik* spaces.

In rural Nepal, negative connotations of forests as primitive, simple, and undeveloped "brush off" even more on forest foragers since they live within such marginal spaces. In Nepal, the concept of the *banvāsi* (*banmānu, jaṅgali*), the Tarzan-like "jungle-man" acts as a metonym for the concept of someone and something primeval. For example, a Kathmandu news reporter interviewed me about the Raute hunter–gatherers and at one point asked, "How many years behind us are the Raute?" As I fumbled to respond that they are not backward or "primitive," he conveyed both concern and pleasure to know of people who could be the target of social upliftment. Farmers, of course, do not view Raute as Raute view themselves.

Raute view themselves as contemporary people who purposely use simple technologies. They consider themselves to be "God's children who live in the forests" as opposed to others who live outside of forested spaces. By situating others as outside of the forests, Raute are, on a deep level of social ordering, calling into question the humanity of farmers and foreigners. It is precisely these more complicated ways that serve as a conveyance for deeper insight into the socio-political relations between Raute and Hindu householders. Places, particularly forests, that have a unique reality for each inhabitant, and more broadly, for each ethnic group or class of people. As heterotopic spaces, forests represent marginal, often jumbled, spaces that have alternate and contested meanings (Tuan 1977; Foucault 1986; Rodman 1992; Owens 2002; Whatmore 2002). And it is in this alternity in which the power to know and name a place develops. Thus while we may state that Raute live in rural wilderness, such places represent entirely different meanings to those moving through the forests themselves. Such heterotopic places are more complex than a simple ethnographic setting; they exist as something at once both an existential concept and a naturalistic object in space. The trick, for those of us who describe such places, is to capture the experience of place and communicate it in one of these dimensions. Thus, I focus not on the decentered analysis of one etic and existential concept of the forest, so much as on the foragers' emic social constructions of forest places. These places have polysemic meanings for different Raute users, since ultimately, forests are physical entities known to us as symbolic and cultural spaces.

Rautes as forest dwellers

Villagers call the nomadic Raute by the names "jungle folk" (*jaṅgali*) and "forest men" (*banmānu*), names that evoke a sense of people who are at once as wild and uncivilized as the forests in which they live. Such designations are not accidental; farmers view Rautes as uncontrolled, wild, sexually charged, and inhabitants of a place of natural wealth of food resources. For example, I asked Maya Khatri, one of my elderly Hindu neighbors in Karkigaun Village, "How are today's Raute and the Raute from before different?" She responded, "They're very different, granddaughter. Back then, we had our ways of getting food and speaking (*launu khānu boli bhāṣā*) and they were just like deer in the forest." Only a handful of elderly farmers remembered calling Raute "Khamshi", which they thought might refer to Raute in the old local Magar language of the region. Some scholars (Singh 1997) also use the name "khamci" to refer to the Raute language using the Raute word for "speech," though Raute themselves do not name their language. They sometimes describe it as *khāmā ghayeko*, which literally means "dancing talk."

Anthropologists, too, have descriptive names for Rautes, referring to them as "hunter–gatherers" since their subsistence strategies are so notable. But Rautes do not reference themselves as hunters. Among themselves, the Raute term for "hunters" is *syela* (alt. *syaola*). They might say, "Hunters are killing the forest animals" (*syela syā mararoka*"). But they don't refer to themselves as *syela* or by the Nepali term for hunters, *sikāri*. Instead, Rautes explain to farmers, using the common language of Nepali, that they are the *bāsasthāne mānche*, the "people who camp or stay here [in forests]." This label conveys the Rautes' sense of self as forest dwellers since they live in the forest and think of themselves as people who are forest dwellers.

Another way that Rautes have described themselves to others is as "the political subjects of the trees" [Nep. *Bot Prajā*] (Nepal-'Yatri' 1983). In this worldview, the forest is their home. They think of it as a politically ordered place rather than as wild and uncivilized. This autonym belies a subordinate relationship to the forest as *prajā* are normally political subjects of a *rājā*, a king. Thus Rautes are implying that the trees are metaphorically not only sentient beings, but politically powerful ones—like kings—while they the Raute are their dependent subjects. As an act of defiance, the Rautes call themselves not simply "*prajā*" in reference to their relation to the Nepalese state, but "Bot *Prajā*" to signal their fealty to the forest rather than the King of Nepal. This represents a significantly different rendering of social space, one that marks the Raute as fundamentally centered in a political space that includes trees as powerful sentient beings. In this context, the label *Bot Prajā* infers that the forest and its trees are one of the most important dimensions of social life for Rautes.

Conversely, Rautes sometimes describe themselves as the "Kings of the Forest" (Nep. *banko rājā* or *banrājā*). This ethnonym, used by *ādivāsi* (indigenous) peoples of India, portrays Rautes in a politically dominant light within the political world of the forest where all creatures are considered as political beings. A polysemic message, the epithet *Kings of the Forest* also connotes a need to compare favorably with other social castes. The reasoning behind this autonym is that, if Raute are indeed *Kings of the Forest*, then their caste status should also be commensurably high, akin to that of the Thakuri caste nobility who live throughout the region. Indeed, Raute, like other foragers who use this autonym, do claim ancient caste affinity with the Thakuri kings and landed nobility. Linguistic evidence, however, points to Raute as being Tibetic-speaking peoples who have been enveloped into the Hindu caste system.

A third way that Raute have described themselves in relation to the forest is as part of the "great family of God." When I asked who is part of this great family, one elder explained that all of God's children live in the forest. For the Raute, the forest is the source of life, of sustenance, where resides all of the great family of God. Raute discussed with me the various roles of the sun, moon, stars, and other non-human supernatural beings that make up their great family. Man Bahadur, for example, claimed that the sun, known as the deity *Berh*, represents their father, and the moon, *Bairha (*alt., *Berha)*, their mother. When Rautes say that they are children of God, they are referring to the creation of the world as made by Berh/Bairha as a creator deity. When Rautes converse with non-Rautes in Nepali, they use the Nepali name, *Bhagvān*, for God. *Bhagvān* is thus the Hindu concept that comes closest to Raute notions of their creator deity. Together Berh and Bairha have had many children, including the forests and the Raute people. Some of the sun and moon's other children are birds, insects, monkeys, trees, stars, tigers, snakes and other various natural and supernatural beings. In this wildly populated sentient landscape, monkeys are especially close to Rautes, being called "our little brothers." For Rautes, not so much a hierarchy but an order of precedence flows from older to younger, men to monkeys, and upriver toward downriver. Propitious directions involve places toward their original home, west of Jajarkot District, and their valorizations of space in general involve forests and river places apart from villagers' sacred places.

Monkeys also figure prominently not just in origin stories, but in many Raute narratives. In another story that reverses the usual sequence of monkey-into-human metamorphosis, macaques and langurs in the beginning used to be just like humans. Indeed, the monkeys even held weddings and celebrations. At one wedding, Raute were invited to a monkey troop's wedding. But during the wedding, a Raute guest accidentally burned the tail of the macaque. This unfortunate event resulted in macaque having its stump tail that today is characteristic of the local species of macaque monkeys. This story was told to me for instructional purposes, to teach the folk-classificatory differences between monkeys and humans. For as the story shows, monkeys have "tails" that are different from those of humans. The Raute storyteller then showed me his "ponytail," (*jyuri*) and pointed out how it was much longer than that of a macaque's little "pigtail" of a tail (*mhutā*).

As another Raute hunter stated, referring to monkeys, "The people *(janatā)* are God's sons and daughters. If they are not God's children, we wouldn't speak about them and God would have to create them. God made them for our help [our food] and they're important."

For most Hindu villagers, on the other hand, monkeys are nothing like brethren; rather they represent creatures of either the natural or supernatural world. They are either worshipped as the deity Hanuman or treated as wild disruptive scavengers that prey upon the corn crops. In the regions north of Raute territories, the primatologists John Bishop and Naomi Bishop found that Buddhist farmers were loathe to hunt langurs since killing other sentient beings is immoral (Bishop and Bishop 1978). Farmers will chase the animals when their grain crops are raided, but normally they refuse to kill the animals. Thus various castes and ethnic groups in the area view monkeys as deities, as sentient beings, as disruptive scavengers.

The great family of God that lives in the forest is not limited by a geography of temporal and physical constraints. Some of their most important relatives are those who are invisible, having died and are now invisible spirits, known by the Raute term *hori* or *hor*. The notion of *hor* is complex, an idea involving the spirits of ancestors. As I understand it, *hor* refers to the deified human spirit that pervades all forest life. In Mundic language groups, *hor* is an important concept that refers to "man" in the sense of humanity, and in a narrower sense to Mundic people and their ancestors (Bhattacharya 1970). Among the Chepang, Tibetic-speaking post-foragers, *hor* is defined as a genealogy, a line of descent, a clan (Caughley 2000: 284). Rautes see the forest as a place of great reverence because it is a location imbued with the "*hori*" atoms of spirit or sentience derived from all the animals, the sentient trees, and their own deceased relatives. As one elder explained,

> You are also *hor*, and God of the forest is also *hor* and we are also *hor*. If God does true then it will be true. If God does false then the whole forest will be destroyed. All Gods are *hori*, because we cannot see them.

In this utterance, *hor* means something like "spirit" and one which is invisible yet omnipresent, powerful, fallible, and found in both natural and supernatural realms. In other interviews, *Hor* has a pervasive quality, like atoms of spirit or sentience that pervade and surround all beings. Broadly, *hor* refers in a mass-noun fashion to the Raute's deceased ancestors, whom they believe exist as spirits, or atoms of sentience, existing in surrounding spaces as Rautes move through the secular world. Rautes move through their mundane forest world at the same time that they move through a morally distinct domain, a sacred world filled with ancestral spirits. This large family of humans, animals, spirits, and deities may have different duties and characteristics, but all are important within their forest home. As one Raute elder insisted, he is no better or worse than any other of God's children.

Occasionally, Rautes communicate the importance of the forest to villagers by delivering to villagers such pithy remarks as "Without the forest, there is no life." Such phrases are heard occasionally in environmental sloganeering in India (Narayanan 1997). It certainly is not lost upon Rautes that their slogans may result in benefits to their group such as protection from political elites like District Forestry Officers. Mayn Bahadur Raute, who said this during a discussion in the bazaar of Jajarkot District in western Nepal, wanted to emphasize that Rautes utterly depend on the forest, but so do the villagers, and that both should respect the forest as the source of all life. Regularly, Rautes try to communicate their dependence on the forest as a place of life and not just a place to be utilized by humans. For the Raute, the forest is a source of ultimate sustenance in both a religious and practical fashion.

The Rautes' ontological domain of domesticity is clearly bounded, however. Places that are not part of where "God's family lives" lie outside the forest, in the domain of villages. Villagers are clearly not part of the Raute "family" in the cosmological sense because they do not live in the forest. They represent a "foreign" yet necessary and interdependent realm in which egalitarian familial relationships do not apply.

Living in the forest: a Raute camp

For Rautes, forests evoke something comforting, familial, and divine. Yet, it is not accurate to say that Rautes live in the forests *per se*, but rather that they live in huts (*nao*) that are gathered together in *camps (naolā)*, and that these camps are just as culturally manufactured and distinct from the surrounding forests as are villages. The camp contains marked boundaries, etched pathways, small drainage ditches, and other physical constructions. Yet because huts are composed of woven *bauhenia* leaves and other simply woven materials, similar to a Native American *wickiup*, villagers gloss Raute camps as "natural" rather than "cultural" edifices. To most villagers, then, Rautes live "in the forest" rather than "in camps" since camps neither look nor function like villages. This spatial category called "camp," imbued as it is with the shaping of materials from nature, marks just one of many important distinctions between Rautes and Nepali villagers.

Villagers want sturdy, two-story plastered edifices that house livestock, family, and a front courtyard for entertaining. They want something enduring, to return home to after adventures in India or Kathmandu. For villagers, their homes represent one's place in the community. Rautes, on the other hand, carry their entire household literally on their backs. They want something ephemeral, replaceable, and in fact, will destroy their *nao* houses in a day's notice before moving on to build their next campsite. One represents an attitude of permanence; the other, of impermanence.

When I visited the Raute camp in Tarapani forest, the 32 huts looked like leafy mounds from a distance. They were set up on the mountain ridge, next to an abandoned goatherd's shack. As I walked on the winding footpaths that led toward the camp, I mentally divided the space into 'zones' based on changes in the forest. In 'Zone 3,' from about 2 kilometers away from the camp, I noticed signs of Raute activity. I heard a ringing *twããk twããk* of their axes as they cut trees. I met Raute women walking in the brush collecting fuel, pine needles, and greens. I walked past trees that had notches or gouges in them. The trunks of large trees had been cut to remove material to make bowls and lids. Smaller trees had their inner bark stripped off. In 'Zone 2', just a few hundred meters from the camp, I came across more numerous signs of activity. Little boys played at cutting wood with imaginary axes. Little girls built pretend huts, trying to drive miniature wooden pole stakes into the ground. Thousands of pale wood chips lay scattered on piles on the ground. Broken wooden bowls lay tossed on the wayside, and even human feces had to be ignored along the way. And finally I reached the edges of 'Zone 1', the camp itself, where human activity unfolded at a slow pace. Elderly women chatted in small groups, a woman stirred a cooking pot of roasting grain, children danced about, a man ambled into camp with his daughter perched upon his shoulder. Off to one side, a young man patiently shaped a wooden breadboard, propping it up on a wooden stand as the chips surrounded his workstation.

As the afternoon approached, I visited the Raute elder Man Bahadur's hut nestled on a bluff 5000 or 6000 feet above a tributary of the Bheri River. I noticed that the *nao* hut was small for a family of four. It measured about 5 by 8 feet and was surrounded by a shallow ditch designed to shunt rain-water away from the interior. His hut led to others along dirt paths made naturally by the frequent tread of relatives from hut to hut. Next to him lived his son-in-law and his daughter in their own hut. Man Bahadur's hut was roofed by leaves woven into a tarp (*lhāpa*), made from a shrub called *mālu* (*Bauhenia vahlii*). Known in English as Camel's foot, *mālu* is a gigantic climber with tendrils that surround trees, eventually choking them to death. Some rural Nepalese still use *mālu* for making ropes and mats. The new pods and leaves are cooked as vegetables while seeds taste like cashews and are eaten raw, roasted, or dried and fried (Manandhar and Manandhar 2002: 106). In sunny spots, leafy covers of *mālu* woven into big tarps, lay drying in the sun along the periphery of camp. They were used not only to cover huts, but to shield the sides of huts from the sun. Shades of these tarps were pitched next to people working and one Raute woman wrapped up the *lhāpa* and stored it in a hunting net (*jabo*). I walked with Man Bahadur's wife over to a *Bauhenia* tree and stripped some inner bark off, noting that it felt as supple as moist rope. Another woman walked some old leaf fencing to the side of the camp and threw it away. She gathered new *mālu* leaves and wove them into the side of a hut. As one Raute noted, "Everything is made of *mālu*—covers, plates, baskets, walking sticks, rope, our eating bowls—everything."

Just outside the huts, fermented grain beer was stored in the 3-ft-high oval-shaped wooden vessels (*jhum*). This beer was not for special occasions; it was consumed like cider, for occasional drinks during the day when Rautes returned from work in the forest. The wooden bowl construction sites were placed next to the huts too. In elder Man Bahadur's work area, a blanket of blond wood chips covered his artisanal arena. A log [N. *achānu*, woodcutting board] formed the base on which the wooden items were situated to be carved. The bowls-in-progress were propped against a wooden stand (N. *kili*) that balanced the bowl as it was chipped and smoothed. Man Bahadur rested against the wooden chest, his arms crossed and legs extended. He noted,

> Right now this forest camp is clean and fine. But soon I think there will be problems with leeches. It is a pretty damp place here, and a few children have already gotten itchy bites. If you have some medicine, let me know. If the leeches are not too bad, we will stay here one or two weeks, then we will be ready to go to the next camp. Except for the leeches, the camp is fine because the women like the nearby stream that comes out of the mountain. We have dug our *pāni dhāra*, water tap, very close to camp.

Even in 1969 when Terence Bech visited the Raute, they told him that they prefer similar water sources, ones not made by villagers. As I hiked into the Raute camp, an island of their inner social space, I realized that they would soon burn down the camp in preparation of moving to their next camp site. But they don't stay in their camps all day long; soon we would also walk to a nearby village for a barter session. As they move from their camp location to the sedentary village, Raute will adopt a different persona than that of egalitarian forager. They will don the mask of high castes, pretend they are long-separated kin of the Thakuri, the Kshatriya elites that rule the region. Such pretenses among tribal peoples in India have a long-standing history. For example, the Austro-asiatic-speaking Bhumij adopted the vegetarian model of caste behavior and proclaim themselves descended from Kshatriya lineages during their process of Kshatriyaization or Rajputization (Sinha 1962). In a similar vein, Tibetic-speaking Raute tell the villagers they are both are of Kshatriya heritage but that long ago, "You were thrown by God into villages while we were thrown into the forest."

Last camp at Kahanda: the leaf's edge

The last camp that I visited during the research season was near Luwida and Majkot, two villages in western Jajarkot District off a route toward the ancient summer palace of Khas Malla kings who reigned in the fifteenth century. To get to the area that villagers called Tarakhola, meaning 'the distant river', I walked through the forest past a small stone temple about 4 feet wide. On all sides, the unknown sculptor carved intricate ten-pointed stars and other symbols, but their meanings, and the deities the people worshipped hundreds of years ago are forgotten (Fortier 1995: 144–80; Pandey 1997: 639–89). The stone stellae and temple tops rise like fingers stretching out of the ground, presenting vivid reminders that the forests are not just empty stretches between villages but part of a constantly changing political landscape.

While villagers call the forest here Tarakhola, Rautes know it as 'Kahanda.' One Raute youth named Mahabiri noted that Kahanda refers to the edges of the *mālu* leaf. Again, I am struck by the constant references to *mālu*, and in this case, to the specific part of it, the *edge*. I've never even thought much about the edges of a leaf, except as a diagnostic feature. Looking at a *mālu* (*Bauhenia vahlii*) leaf, I noticed that its huge leaves are shaped like a camel's footprint. The leaves are as broad as they are long, with a cleft cutting through the middle. Mahabiri didn't elaborate as to why the forest is called Kahanda, and I neglected to ask about its topogeny, or what story this place name evokes and its relationship to other place names. Does it recount a journey of an ancestor? The migration of their group? Is the name simply a metonym of the forest as the shape of the leaf? Mahabiri only pointed to a nearby *ayār* (*Lyonia ovalifolia*, no English name) tree and said,

> We camped here last time we came through here. See the *ayār*? It was cut down by us twelve years ago. And next to it can you see the spring coming out of the rock? Its open now, water's flowing where we made our drinking spring. We made the new spring nearby that old spring [pointing], but that old one was made by villagers for their cattle.

Mahabiri's observations reminded me that these forests are both new and old for nomadic Rautes. Mahabiri couldn't have remembered this forest more than once in his life since Raute had only traveled this route twice in the past 20 years. Yet he came to Kahanda looking for, and finding, physical impressions made by his own people, the features of sacred cultural landscape that give Rautes the right of usufruct for a short time.

From threatened beliefs to threatened lives

Non-Raute interest in control and use of forests has been omitted up to this point, but it deserves our attention. State interventions do occur and these interactions between Rautes and the Government of Nepal involve power differentials over how to use and view forest spaces. States have interests in forests that affect forest-based peoples, especially in the appropriation of its resources, the regulation, and the taxation of forest timber and non-timber products. Because of the enormous power of governments to affect the lives of their citizens, the edicts and actions of the government of Nepal concerning forests and forest foragers can lead to negative results. During my fieldwork, Raute elders expressed concern over the fighting between Maoists and the Government of Nepal. One Raute elder reportedly requested the Government to be careful during aerial bombings over the forest. When asked if they had been affected by the recent conflict, he replied,

> The sirs who fly the *gādi* (helicopters) should be careful. What if we Rautes are completely wiped out when they are trying to kill Maoists? However, the CDO [Chief District Officer] has assured us that our areas won't be bombed. Just to be sure, when a helicopter flies overhead we wave our turbans to notify the sirs that we are not Maoists but Raute and to leave us in peace.
>
> (Lawoti 2005: 11)

Raute do their best to confront these government actions, often meeting with the Chief District Officers in regions through which they travel. Yet there is always the possibility that one of these officers will persuade Raute to settle down, take government food rations, or even block the nomads from using forest resources. Their ability to avoid becoming a "criminal tribe," or forest peoples who illegally collect forest resources, is contingent upon the good will of individual elite men. This represents a precarious situation and Rautes risk losing the patronage of local elite political figures if they divulge their radically different views and beliefs about forests.

But to what degree are Raute living in a heterotopic space of disparate geography while they dwell in the forests of Nepal? To those who share Foucault's view of heterotopic sites as imbued with power/knowledge, the forests indeed are jumbled spaces immersed in contestations of political power over the right to define, to *know*, geographical places. The perspective of the subaltern is pushed to the side in favor of hegemonic perspectives of place. In the Raute case, we can understand how the politics of naming, knowing, and using natural spaces such as forests plays out. Nepalese villagers, speaking their Nepali language and wielding a righteous Brahmanism, seize advantage over the "broken tribes" of Nepal at every point of contact. The villagers conceive of forests and nature as a heterotopia constituted by divine Hindu bodies, while Raute envision forests as a domestic space created by God/Berh.

Looking at the Rautes' own topophilia, their own affective ties with the material environs, we discover another crowded forest-world inhabited by deities, humans, spirits, and animals-as-relatives. This represents a whole different form of jumbled space, one with a dynamic tension between supernatural and natural beings. Here lies an extended family that Rautes perceive as more like a busy household. When individuals do something immoral, conflict occurs that can destroy their forest home. As Raute elders emphasized, maintaining their cosmo-ecological balance is of prime importance. The Rautes' role is to propitiate Berh, live a morally good life, hunt monkeys, and continue to migrate to new foraging grounds. If Raute act immorally, their world will be destroyed. Berh would make the monkeys disappear, and Raute will lose their taste for monkey meat. The forested world of "God's children" represents a dynamic heterotopia, one where animals and deities have as much agency as their human counterparts.

In the conflicting views between sedentary and nomadic societies, political issues are always paramount. Raute make their bid for cultural autonomy by asserting themselves as subjects of the trees, as kings of the forest, and as nomadic forest dwellers. They evoke a political image of people who wish to have their human rights respected. Even though this is an endangered society, Raute nevertheless assert their right to maintain their own culture, cosmology, and view of nature and the environment. The Raute elder Ratna Bahadur emphasized their right to exist when, using a typical forest-based metaphor, he said, "We're not just the bark of the tree, you know. We're important too."

6 Clear mountains, blurred horizons

Limbu perception of their physical world

T.B. Subba

Introduction

No one who has ever written on the Limbu community has considered their perception of physical environment important enough. The early British writers, including Eden Vansittart (1915: 108–9), E. T. Dalton (1872: 102), H. H. Risley ([1891] 1981: 18–19), C. A. Campbell (1845: 75–8), and John Morris (1985), have written on the history, language, clan organization, religion, and *rites de passage* of Limbus but they literally wrote nothing on the Limbus' relationship with their environment. Among the more recent publications on them, in the Nepali or English language, are dictionaries, books and articles on their relationship with Bahuns (Nepali Hill Brahmins) or on their history and culture (Chemjong 1960; Caplan 1970; Upreti 1976; Senior 1977; Limbu 1989), or language (Subba 2002). Compared to the Limbus of East Nepal, nothing has been written by the British administrators on the Limbus of Darjeeling and Sikkim: the former perhaps because they were considered as one of the Nepali castes and the latter because they lived away from the Indo-Tibetan trade route resulting in their ignorance about the early existence of Limbus in present-day West Sikkim.

The Lepchas and Bhutias, the other two communities native to Sikkim, called them Chong (also spelt as Tsong) and were thus fully aware of the Limbus, and had even enacted a blood 'treaty' with the latter called *Lhomentsongsum*[1] at a place called Kabi Lungchok in North Sikkim in the second half of the seventeenth century. Sometime after Sikkim became the 22nd state of India, *Lhomentsongsum* became *Lhomen* (Bhutia and Lepcha) for all practical purposes, as the political distance between Bhutia-Lepchas and Limbus widened. The former two communities were recognized as Scheduled Tribes and their seats in the state assembly were reserved whereas the lone Limbu seat was de-reserved and they failed to get constitutional recognition as a Scheduled Tribe until December 2002.

Every Western writer on the Limbus has described them as a 'tribe'. Once a community is so labelled, its relationship with nature is taken for granted. The tribes are conceived almost everywhere as people with a symbiotic relationship with nature. Such is the dominant perception at least in Indian anthropology. Although this could be true for a large number of tribes in India, there are significant variations among tribes in respect of their relationship with forest/nature across regions and even within a region. The tribes today occupy different levels of development, pursue different vocations, and hold absolute political power in some states of Northeast India. Yet, while their way of life has changed, often very drastically, anthropological constructions of them have not.

A corollary of such perception is the widespread belief that tribes are repositories of traditional or indigenous knowledge, especially regarding management of natural resources. There is much evidence in Northeast India to show that they have remarkable traditional wisdom relating to the management of natural resources (see, for example, Agarwal and Narain 1999; Aier and Changkija 2003: 333–79). This evidence relates to management of natural resources by various communities living in the region on the basis of their traditional knowledge, although at times such knowledge has been essentialized, if not romanticized as well. Such knowledge actually depends on the degree of engagement a culture has with natural resources and not whether one belongs to a tribe, and can vary significantly even within a family. For instance, my mother, who collected fodder or firewood from a nearby forest every morning until she was quite old, had far more extensive knowledge about plants than my father, who taught in a school in Bhutan. Once a person's engagement with forests is over, his or her knowledge of the forest inevitably erodes, and it does not seem to continue with each subsequent generation as many other aspects of culture like language, food habit, and dress often do. The very presence of a large number of rituals among tribes everywhere to seek blessings of the supernatural powers for a bumper harvest and good health of people and cattle in the coming year indicates how vulnerable they must have been to the vagaries of nature or how inadequate their wisdom to ensure good harvest and health for themselves.

Many Limbus are reasonably knowledgeable about the management of land, forest and water, although they, like other communities living in the middle hills of East Nepal and West Sikkim, have failed to protect their forests from being denuded or their wells and springs from drying up. It appears that they underwent a major shift from shifting to settled or terraced mode of cultivation after the political unification of Nepal in the 1770s, and such a process continued in some parts of these Himalayas until quite recently. Not only were the areas of shifting cultivation in Nepal brought under settled agriculture, even primary forests were reclaimed for agricultural cultivation to meet the food needs of the population that was growing rapidly due to the arrival of high-caste migrants from the western part of Nepal. Similar changes took place in the hills of Darjeeling and Sikkim. There was a marked emphasis in Darjeeling on tea and cinchona plantations, construction of roads and bungalows, etc. Land was also reclaimed for agricultural purposes but that was more or less confined to the Kalimpong subdivision of the district bordering Bhutan. In the absence of concepts like sacred forests, community forests or clan forests, which exist in some parts of Northeast India, nothing stopped the Limbus from clearing the forests, not even the powerful forest deity, Tampungma.[2] With the decline in shifting cultivation, at least from the latter half of the nineteenth century, their knowledge related to this age-old agricultural practice also began to erode. It appears that Limbus combined this mode of cultivation with pastoralism and trade for a very long period in their history.

In this chapter, I argue, with the help of ethnographic materials based on the Limbus of the eastern Himalaya, that Limbu perception of the physical environment is intermeshed with their social and spiritual worlds.[3]

Who are the Limbus?

I have answered this question in my earlier writings without any hesitation (see Subba 1990, 1999a:180–9; 1999b). It was actually quite a simple question for me then. But now, as I ask myself this question again, I am acutely aware of the difficulties involved in defining who the Limbus are. This is perhaps due to the fact that the community started championing certain symbols of its identity in different parts of the eastern Himalayas from 1990, following the restoration of democracy in Nepal. When I first began my fieldwork on the Limbus of East Nepal in the winter of 1992, they were all obsessed with the need to showcase something of their own culture and tradition as members of other *janajāti*[4] communities were doing. There was neither time nor inclination to debate the various symbols of Limbu identity then. But now a rethinking has begun. One factor necessitating such a rethinking has been the lack of clear boundaries, particularly between Limbus on the one hand and Rais, Yakkhas,[5] and Lepchas on the other. How Limbus relate to these communities is important because they have been neighbours for ages. Let me dwell on this point briefly.

First, the various British authors mentioned above have rightly pointed out that Limbus have traditionally intermarried not only with the Rais and Yakkhas living west of the Arun river but also with the Lepchas living to its east. Such intermarriages are not really considered as 'inter-marriages' in the sense of a marriage between a Magar and a Gurung, and have led to blurring of the boundaries, among these communities, and indeed there is a great deal of similarity in their ways of living. There is, equally, a great deal of compatibility even in ritual and spiritual matters among them, given that a Limbu priest can officiate in a Rai or Yakkha ritual and vice versa. Of these four groups, the influence of Hinduism is most visible among Rais, who, being located west of the Arun, were exposed more and earlier than Limbus or Lepchas living east of this river. The spread of Hinduism in Nepal and the Sikkim Himalaya has been from the west to the east, and that of Buddhism from the north to the south. The long malarial *tarāi* [flat land connecting the plains with the hills] made it impossible for the Hindus to move from south to north.

Second, because they live under similar ecological conditions in the middle hills, the material culture of the Limbus is similar to that of the Rais, Yakkhas and Lepchas in terms of settlement pattern, house types, and food habits. For instance, pork is not only an important item of gift in marriage but is also an equally important sacrificial item for all these communities in East Nepal and Sikkim. So is fermented millet and rice-beer, which are also essential to welcome guests and show respect to the elderly. The practice of offering fermented millet in a bamboo container, which is sipped with the help of a slim bamboo pipe, is seen in the marriage and funeral ceremonies of all these communities. Even the vessels for cooking and storing, worshipping of the main pillar at home or the deities of hearth and cowshed are quite similar. No one can really tell on the basis of such cultural practices whether they belong to a Limbu, Rai, Lepcha or a Yakkha. They all offer a little bit of new crops to the fire or to the spirits wandering around before they start to eat the same themselves. The Lepchas of Sikkim are, however, quite clearly distinguishable from their counterparts in Illam district of East Nepal by their dress and religious practices (Gorer 1984: Chapter 7).

Third, animism as it is practised by Limbus is based on a body of oral traditions they call *mundhum*. Actually, variations in its pronunciation like *muddum*, *mundhum*, and *muddhum* are quite common throughout eastern Nepal and it is indeed difficult to associate a particular pronunciation with a particular ethnic group. One of the best accounts of *muddum*, as it exists among the Mewahang Rais of East Nepal, is by Martin Gaenszle (1993). In this article, he describes how this repository of oral traditions has changed through interaction with written traditions alienating the people from their own past.

These oral traditions are basically accounts of the origin of various flora, fauna, clans, routes of migration, and procedures of conducting *rites de passage*. Recitation or interpretation of a particular *mundhum* by a shaman while performing a ritual is sometimes contested. Elderly persons assembled on such occasions listen to the deliberations of the shaman carefully, and are quick to point out a mistake or remind the shaman of a certain detail left out by him. A shaman rarely objects to such interventions, although at times serious arguments follow between him (rarely her) and elderly persons either after the ritual is over or when the shaman halts the ritual for a break to sip his *tongba*.[6] The use of fowls, pigeons, pigs, eggs, flowers, grains, water, etc. as sacrificial items is also common among these animistic communities, although there are sects cutting across ethnic boundaries that are purely vegetarian and non-alcoholic. Absence of a permanent place of worship is also notable among the followers of this religion, suggesting a common past characterized by pastoralism and shifting cultivation. Animism is, in fact, not just a religion for them: it is also an interactive relationship with their physical environment and with the world of gods, goddesses, spirits, and ghosts whose response to the deeds and intents of the people is almost instant. The supernatural beings reward as well as punish people, help them find lost jewellery or people, and identify the particular 'being' responsible for a sickness or calamity. They bring great crisis when they are angry and yet are forgiving when pardon is asked for with appropriate sacrifice and ritual. They are also compromising with a given situation and accept a fowl instead of a goat, an egg in lieu of a fowl, a leg in place of one whole animal, etc.

Fourth, in terms of physical characteristics it is equally impossible to separate Limbus from Yakkhas, Rais and Lepchas, and one finds a wide range of physiological variations among them, even among Limbus. But there is a stereotypical Limbu face in the minds of people, and when a Limbu does not bear those features, he or she is often a subject of gossip or ridicule. I have heard small children who did not look like a Limbu being asked about their 'real' (read biological) father, who only their mother would know.

Fifth, even when one considers women's ornaments, tattoos and dress it is not possible to make a distinction between a Limbu and a Yakkha woman, or between a Yakkha and a Rai woman. Quite frequently a woman in a Yakkha family is actually a Limbu and a woman in a Limbu family a Lepcha before marriage.

Finally, Limbus share certain clans like Youngmu, Sitling, Lucksom and Mangmu with Lepchas. These clans are considered Lepcha if they are in Lepcha country and Limbu if they are in a Limbu area. They have similar taboos and totems, similar concepts of gods and goddesses reigning over their forests, rivers, mountains, agricultural fields, cow sheds, central pillar of their houses, lightening, hails, and so on. As a matter of fact, if they live in the same village they can understand the dialects spoken by each other fairly well. This is also true of Rais belonging to various dialectical groups living in the same village: the much-hyped mutual unintelligibility of Rai dialects needs unpacking at the earliest.

The identity of Limbus, as of many other Tibeto-Burman language-speaking communities in the eastern Himalaya, has become a very complex issue today, not only due to differences in the way Limbus construct their past but also due to the influences of Hindu, Buddhist and Christian religions during the past two centuries. In other words, it is difficult today to define Limbus in terms of any objective criteria, just as it is perhaps with regard to other, analogous communities. Despite the effort of every community to maintain a measure of internal homogeneity, the external forces that undermine the forces of homogeneity are apparently far more successful in creating greater heterogeneity among them. The people, however, have some idea – no matter how vague – about what it means to be a Limbu, although such a stereotyped concept is primordial in the sense that they are based on language, religion, and culture rather than a subjective identification. But with increasing heterogeneity, the only acceptable basis for defining them may soon be some subjective criterion: they are Limbus because they believe themselves to be so and/or they are recognized by others to be so.

Economic history of the Limbus

There are several indications to show that Limbus were once pastoralists and they probably took their cattle to higher altitudes in summer and came down to the middle hills during winter. One of the clearest indications of this is their categorization of the seasons into two: *songmang* and *thangmang*. The former refers to a movement towards lower altitudes and the latter indicates a movement in the opposite direction. These opposite movements – descending and ascending – are characterized by many changes in climate, flora, fauna, and corresponding restrictions or relaxations of the rules on what people can do or eat. The starting of the two seasons is characterized by sacrifice of chickens in the name of Nahangma, the goddess of power. Apparently, they had quite an intimate knowledge of Tibet, perhaps because Tibetans came down during winter to pay tribute to Limbu chiefs or landlords (Sagant 1996: 423). The other indication is that there are several references to the high altitude flora and fauna in their *mundhums*, including the yak (*Bos frontalis*), which is not normally found in the middle hills where they have been living for the past several centuries. The bushy tail of a yak is considered auspicious and is kept as a precious item in almost every Limbu house. The name Yakthungba, which they use to refer to themselves, also means 'yak herders', which indicates that they were probably yak herders at some earlier time in their history.

Pastoralism is not practised any more, certainly not on the scale and in the manner of the past, although domestication of animals and birds, particularly for the purpose of traction, milk, sale or sacrifice is quite common. There is hardly any Limbu household without some fowls, one or two pigs, a couple of goats or sheep, and cows, oxen or buffaloes among the larger animals. The richer or larger households often have a large number of such animals, which are left free to graze on agricultural fields after harvest in winter or are taken to nearby forests for grazing. Where some vegetables are grown in the winter, such areas are temporarily fenced with bamboo splits, which the animals sometimes break to eat the crops. Such incidents may lead to serious altercations between the owners of the cattle and the vegetable gardens, but are settled with a token fine and some bottles of liquor, which is soon distributed among the adults gathered for the purpose of settling the disputes.

Until recently, Limbus traded raw wool from domesticated sheep, coarse blankets woven from it, long and sturdy hair of pigs, bear bile, glass beads, aluminium/brass/zinc utensils and cattle. Such traders moved in groups of two or four, for three was considered as inauspicious number. They carried long *khukuri*s[7] to protect themselves from dacoits (*ḍakaitī*, bandit) when they passed through uninhabited mountains or deep forests. They carried their own rations and rarely accepted hospitality of any stranger, lest they were drugged and looted. Several such men earlier visited the hills of Darjeeling and Sikkim selling utensils and buying back old and unusable utensils at a lower cost. A few still do so in winter. They also visited these hills to transport oranges from orchards located in remote villages to the nearest market until the orchards were destroyed by disease.

Further, there are indications to show that Limbus were once shifting cultivators. This is first indicated by the fact that they had communal landownership called *kipat* until the middle of the twentieth century (Caplan 1970). In fact, shifting cultivation and communal ownership are closely associated even in the hill areas of India's Northeast, where privatization of landed property is now taking place as rapidly as the decline in the acreage under shifting cultivation. Even the decline in the Limbu practice of shifting cultivation coincided with the introduction of permanent or terrace cultivation by the Bahuns and Chhetris from the west in the aftermath of the political unification of Nepal. According to one travelogue (Das [1902] 1970: 3), shifting cultivation was practised by Limbus until the end of nineteenth century, although the rotation cycle had come down to three years. Further, worship of the forest deity Tampungma indicates their heavy reliance on forest in the past. The forest deity is believed to look like an old woman with grey hair carrying a basket and a small sickle. Limbus consider this deity to be very important for their overall prosperity. That they might have once been shifting cultivators is also indicated by the fact that they rarely have a permanent place of worship. They create an altar for worship at any clean place by slightly raising the platform and decorating it with plants like bamboos and plantain trees. For certain rituals the decoration of the place of worship is extremely elaborate but everything is dismantled at the end of the ritual and thrown into a stream or left at a clean place. No trace of the ritual performed or the articles used remains. The next time they have to perform such a ritual they find some other place, although sometimes there are restrictions related to the orientation from the house for performing a ritual.

In fact, both pastoralism and shifting cultivation give rise to a non-sedentary life, which is reflected in the deities they worship, the rituals they perform, the places of worship, the materials used for all this, etc. Such a life also depends on collecting food from forests and rivers rather than storing it for the lean season. Although Limbus have been sedentary for over two centuries now, the traces of their non-sedentary life can still be seen. Even today, a Limbu will leave his village and go to another on the slightest pretext.

Their physical environment

The physical environment in the eastern Himalaya has been steadily degraded in every respect. The forests have vanished, wells and streams are dry during winter, and the soil has suffered erosion. The herbs that Limbus have used for treating diarrhoea, headache, nausea, skin diseases, etc. are no longer available nor do they easily get the plants they used for their rituals. Their crops fail frequently and are subject to the depredation of wild birds and animals. Landslides are frequent and sometimes hazardous to humans and cattle.

Almost everywhere in East Nepal, West Sikkim and Darjeeling, Limbus are predominantly settled in rural areas of the middle hills at altitudes varying from about 3000 to 6000 feet (approximately 915 to 1830 metres) above sea level. The climate is temperate and the vegetation predominantly tropical. If the soil is fertile, depending on the altitude, slope and aspect of the hill, they can grow many varieties of crops, both for domestic consumption and for sale in the market. Marketing crops is an opportunity rarely available to most Limbus in East Nepal and West Sikkim due to the distance between their villages and the nearest market. Therefore, they generally grow maize as a summer crop and paddy and black lentil as winter crops in the irrigated fields that are normally limited to 1500 metres above sea level. In the winter the high altitude dwellers – Sherpas and Bhutias – come down to the middle hills and exchange their crops with Limbus, Rais and Yakkhas.

In the dry fields, which may or may not be terraced, they grow either maize, black lentil or millet. Inter-cropping with beans, chillies, yam, pumpkin, tapioca, turmeric, sweet potatoes, arrowroots, etc. is common, which also indicates a shifting cultivation past. Although shifting cultivation is no longer practised, the cropping pattern is similar to what it was under that system. Almost every available piece of land is used for growing some crop, leaving little room for trees. Even the bamboo groves that grew everywhere in the past are no longer easily seen. Limbus do not encourage the cultivation of bamboos because the land around bamboo groves becomes infertile.

Yet they have an intimate relationship with bamboo. They need it for constructing their houses, cow sheds, pig sties, etc.; fencing their kitchen gardens, providing support to inclining plantain trees, as handles of various agricultural implements including the axe, for making baskets to carry as well as to store grains, as pipes for drawing water from long distances, harvesting fruits and crops from tall trees, among other uses. They also need a particular variety of it for making *tongba* (see note 6). They need green bamboo for carrying a corpse to the cremation or burial ground. Its leaves are a favourite for cattle as fodder; its shoots make good curry and pickle; its roots provide very good hockey sticks, tops, and other playthings; when dry it burns very well and also helps to burn the less dry firewood. Limbus do not use green bamboo to cook rice, as many tribes in India's North-east do occasionally, but make various kinds of handles and weapons out of it, most importantly bows and arrows, and handles for *khukuri*, spears, and sickles. Best of all, bamboo does not flower and create famine by contributing to the extraordinary growth of the rat population, as it does in Mizoram, one of the states of Northeast India. But Limbus believe that they should not take shelter in a bamboo grove at night. Doing so is believed to be hazardous because ghosts and spirits are attracted to such groves in search of victims.

The other important plant is the plantain. Traditionally, most Limbus rarely eat any fruits except during festivals like *Dasaĩ*[8] and *Tihar*,[9] which were boycotted in the 1990s by many Limbus who claimed that they are actually Hindu festivals whereas the Limbus are not Hindus (see Subba 1999b). They sell ripe bananas in the market but not more than a couple of times in a year. They do not eat the tendermost part of the banana trunk like many other people in India do. They have dishes based on the banana flower, but only on special occasions such as marriages and festivals. But they always keep a plantain grove near their kitchen garden. The reasons are many. First, the banana leaves are used for sacrificial purposes. They make small containers, lids, and plates out of banana leaves on which they offer meat, eggs, rice, coins, etc. at the time of sacrifices. They also erect the banana trees at the temporary altar they make for various rituals. The banana leaves are used as an inner covering in aluminium pots when they boil such things as millet, tapioca, sweet potatoes and yams; the leaves provide excellent air pressure and the articles are boiled evenly and faster than when not covered with such leaves. When banana leaves are not available, fig leaves are used, which are equally effective, but not as acceptable as banana leaves for ritual purposes.

Not much fauna remains today. With the forests gone there are hardly any wild animals left. Even the monkeys that were once a big nuisance to crops have vanished along with the forest. Half a century ago the situation was quite different. There were primary forests along riverbanks or gorges where bears, leopards, monkeys, boar, and deer were available to hunt. Porcupines and wild fowl were also easily available. But today there are virtually no wild animals even where there is re-afforestation, as the new forest consists of monocultures that are fast growing but with little or no social or ecological value. These trees are meant for industrial use and they do not sustain wild animals or birds. The disappearance of birds from their physical surrounding has had a very significant impact on agricultural production as well. With no birds, no butterflies, no bees, and no beetles around, pollination is badly affected. The crops grow very well with the help of chemical fertilizers but the yield is often disappointing. The spraying of DDT powder, which is done not only as malaria-control measure but also to get rid of the insects afflicting the crops, has also caused the disappearance of the pollinators.

Rats are some of the smallest creatures Limbus interact with. The rats not only eat the stored grain, but they often change the course of irrigation and even cause the terraces to break by digging holes through which water gushes out. Unlike some of the tribes of North-east India, Limbus do not eat rats, not for any taboo or totem but perhaps because they are a proud people. They may borrow but will not beg. They might be poor but they are rarely prepared to plead with high caste authorities to extend benefits to them. Limbu women are even more notoriously proud. 'Why would Limbus marry so many Lepcha women?' I was asked in 1992 by an old Limbu in Yangnam, Panchthar district of East Nepal, when he came to know that my wife is a Lepcha. Before I could find a sensible reply, he answered himself saying, 'Because Limbu women are very proud and dominant.' This old Limbu whose own wife was a Limbu, also had a theory to explain why the Yakkhas lost their identity. According to him it was because they married too many Limbu women. Such a proud people have an interesting relationship with rats.

Although Limbus often use traps and poison to kill rats, they also believe that to have rats at home is a sign of a wealthy person. If there are no rats at home that means there is nothing to eat; the family is poor. In other words, although a nuisance in the terraces, the rats are also a symbol of prestige. Further, when a child's tooth falls or is extracted it is offered to the rats in the house saying that they take the tooth and give a new tooth, strong like that of a rat, in return.

Limbu perception of their physical environment

This issue has been extensively addressed in the previous section. My theoretical position, as stated already, is that the concept of their physical environment is inseparable from their social and spiritual worlds. Similarly, Aggarwal and Russell also show (in Chapters 5 and 3, respectively, in this volume), with reference to Kumaun and a Yakkha village in East Nepal, how the sacred and the profane worlds of people overlap.

In order to demonstrate my point, I first list some environment-related concepts in the Limbu language. Although all Limbus can speak Nepali, I exclude the Nepali concepts from the purview of discussion here on the ground that it is a quite recently acquired language. Use of Nepali words would also make the Limbu perception indistinguishable from Nepali perceptions. Some of the Limbu words related to environment, directly or indirectly, are as follows:

Limbu	English
Cha sot	the offering of the first fruits of harvest to the household goddess, Yuma
Dung dunge	god of the Samba clan; the daughter inherits the god's worship from her mother and propitiates him with the sacrifice of an uncastrated goat
Hang pha	a species of bamboo used for making *tongba* or the container in which fermented millet is poured and sipped with the help of a bamboo pipe
Hang sitlang	the main post of the house where Oama, or the household deity, resides and brings luck and prosperity to the household
Keba	the tiger spirit that eats the liver and heart of its victim
Kebo	the spirit of the womb responsible for difficult childbirth
Kham	land, soil, earth
Khamplep	earth, soil
Khim/Him	house
Lam sakma	closing of the path of the forest spirit by the shaman after driving it away from one's body or house
Lekwa	clay, mud

Lising pha	a sturdy variety of bamboo common in the middle hills, appropriate for construction of houses
Lok pha	a variety of bamboo used for making hunting instruments
Makhyu	bear
Maki	maize
Mendohi kokenamba	god of the soil who is paid for the burial land
Miyong Shire	spirit of the wild cat that is fond of eating the livers and hearts of its victims
Mutlama	a variety of bamboo
Nahangma	a female deity that resides on the top of snow-clad mountains, who is worshipped for regaining power (*Na* meaning 'hereafter', 'otherwordly', *hang* meaning 'power', and *ma* meaning 'woman'). She is considered a warrior goddess and is offered a sacrifice of chicken twice every year, during the descending and ascending seasons called *songmang* and *thangmang*
Nam	sun
Nam get	rising sun, east
Nam tha	setting sun, west
Nam yoba	lavender, useful for purification rites
Ningma phuma	the creator of the world, the omni-present, male god
Pha	bamboo
Pha yet	a bamboo winnower
Phung sam	'tree', 'flower' or external soul
Pitkekomba	a cattle-herder, who grazes cows and buffaloes
Pung punge	spirit of springs and waterfalls
Saba sam	monkey spirit whose path must be closed before making any sacrifice at home
Samyok lung thim	the evergreen grass and stone used for purification ritual
Sijongwa	pure (rain) water used for certain rituals
Sing	tree
Songmang thangmang	declining season, rising season
Sung khimba	an aromatic plant used for treating malaria
Tampung	forest
Toksongpa	god of ridges or mountains
Wabung	source of a river
Warok	lake
Warokma	goddess of water
Ya	rice
Yakhe	yam
Yumikma	the lake of souls

(Sagant 1996, glossary)

From the above select list of environment-related Limbu words, one can draw the following tentative conclusions. First, the variety of bamboos mentioned in their language indicates that they were once a pastoral community, a point I have tried to make earlier in this chapter, as they were familiar with bamboos that grow in the high mountains as well as those that grow in the middle hills. Second, the physical environment is not just physical but also spiritual, as various gods, goddesses, spirits, and deities reside in the forests, streams, waterfalls, mountains, animals, and most importantly human beings as well. Human beings do not apparently occupy any superior position in relation to the physical and spiritual worlds; rather their health, prosperity, etc. are dependent on the pleasure of the supernatural beings. Third, even the houses are not free from the deities like Yuma and Kapoba (literally grandmother and old man) and the main pillar of the house is sacred like the hearth and the granary are. Fourth, the above glossary clearly shows that animism is their religion, although it is common between the Hindu and Buddhist Nepalis to share some of the characteristics of animism. Finally, they do not have a single word for the environment.

I think the concept of animism, as it is practised by Limbus in the eastern Himalaya, calls for some discussion at this juncture because it not only promises to unravel the overlap between the human, the sacred and the physical worlds among Limbus, but it might also hold the key to understanding how they conceive of their physical environment. Animism forms the sub-stratum of Limbu beliefs even as Hinduism and Christianity heavily influence many of them. As a matter of fact, this is also true of most tribes of Northeast India who have taken to Christianity.

I begin the discussion on Limbu animism by quoting some early British writers who, I think, are fairly accurate in their description of the Limbu religion. C. A. Campbell, one of the earliest to write on the Limbu religion, says that they are neither Hindus nor Buddhists and it is indeed difficult to give a name to the religion they practise. He writes that Limbus believe in an all-powerful god but do not build temples. To quote him:

> The usual form of worship consists in making small offerings of grain, vegetables, and sugarcane, and in sacrificing cows, buffaloes, pigs, fowls, sheep, and goats, to one, or any and all of the gods, and in eating the flesh afterwards, or as it is pithily expressed by themselves in dedicating 'the life breath to the gods, the flesh to ourselves'. The usual places of sacrifices are merely marked by the erection of bamboo poles, to which rags previously consecrated, by having been offered up, are tied; these are generally placed for convenience at the road sides, and a cairn of stones collected at their base.
>
> (Campbell 1845: 77)

Edward Dalton, who wrote less than 30 years later, draws the section on Limbu religion almost entirely from Campbell. Similarly Eden Vansittart, who wrote in the early twentieth century, relies on the authority of Herbert Risley and Sarat Chandra Das. While the latter's classification of Limbu priests is widely quoted by subsequent writers, this is not a correct representation. Das's main flaw is that he has confused the various sacred/medical specialists whom Limbus consult for various purposes with Limbu priests themselves. To me, Risley provides a more accurate and exhaustive account of Limbu animism and argues at length that it is not really shamanism. He is at the most willing to concede it only the status of elementary shamanism. He writes:

> [T]he Limbu religion may be defined as a rather elementary form of Shamanistic animism, in which the Bijua and Phedangma play the part of Shaman, the former operating on the demons, and the latter having for his department the gods. Finally, we may perhaps hazard the conjecture that the original religion of the Limbus is closely akin to the Pon or ancient religion of Tibet. In both we find the forces of nature and the spirits of departed men exalted into objects of worship. In both systems temples and images are unknown, while propitiatory offerings occupy a prominent place. To complete the parallel, neither recognize(s) a definite priestly order, while both encourage resort to Shamans or medicine men to ward off the malign influences which surround the human race.
>
> (Risley [1891] 1981: 19)

Let me make a few comments on Risley before continuing. First, the division of labour for Bijua and Phedangma is not correct even historically. These shamans are called in to ward off any crisis that a Limbu thinks is connected to the supernatural world. Second, the relationship between Bon or Pon and Limbu animism is interesting and worth exploring a little further, which I do a little later, but what he says about spirits of departed men is certainly mistaken. They are (and were) propitiated with offering only if a Phedangma's divination shows that such a spirit is still wandering around unhappy and unfulfilled and is the cause of sickness of someone in the family. It is, however, true that there is no temple and no worship of idols in Limbu animism, although of late they have constructed one temple in Sikkim, which was a matter of much debate among Limbus for some time.

The similarity between Pon, which is a pre-Buddhist religion of the Tibetans and Bhutias (Tulku 1991), and Limbu animism is striking. This is quite clear from the following list of overlapping concepts, although the Limbu and Tibetan words may not be same or even similar (key: L for Limbu and T for Tibetan):

External (tree/flower) soul:	*Phung sam* (L)	*bla-gnas or bla-sing* (T)
Lake of souls:	*Yumikma* (L)	*bla mtsho* (T)
To 'hold the head high':	*Sam phongma* (L)	*dbu-'phangs mtho* or *dbu-rmog* (T)

Slave:	*Yog* (L)	*g'yog-po* (T)
Gods of the land/soil:	*Okwanama* (L)	*gzhi-bdag, sa-bdag* (T)
Power, king:	*Hang* (L)	*dbang* (T)
Carcass of pig (necessary for marriage):	*Khokpa* (L)	*khog-pa* (T)
God of women:	*Manguenna* (L)	*mo-lha* (T)
Most important soul of man that resides on top of the head:	*Mukuma sam* (L)	*dbang* (T)
Internal pollution as a result of internal quarrels:	*Mangde, Nahen* (L)	*nang-dme* (T)
Household deity that dwells on the central post or fireplace:	*Oama* (L)	*nang-lha, thab-lha* (T)
Pig:	*Phak* (L)	*Phak* (T)
Lifting of the head:	*Sam phangma* (L)	*dbu'phangs mtho* (T)
High, up:	*Tho* (L)	*mtho* (T)

(Sagant 1996, glossary)

I deal here briefly with *Sam phongma* (to 'hold the head high') to describe a dimension of Limbu culture that links them quite clearly with the pre-Buddhist culture of Tibet, which also was animistic. This also seems to have some connection with the exaggerated sense of pride the Limbus have about which I have briefly commented above. This is normally a problem only with the male head of a household. When he falls sick and there is no apparent reason for his or her sickness and divination by a Phedangma does not lead to any useful revelation, a ritual to 'lift the head up' is performed. The indications of a man whose head has, so to say, 'fallen' are generally as follows. He remains quiet and withdrawn, does not eat normally, does not sleep normally, has no apparent bodily ailment, and there is no brightness in the face. There are various causes of such a state of social or psychological morbidity. For example, he may have been insulted, assaulted, or abused by someone junior in kinship relationship; he may have failed to fulfil a promise made publicly or privately; a younger sibling or cousin may have married earlier than he; or he may have suffered the exposure of adultery by his wife. All this indicates some kind of dent in the person's ego, pride, or conscience. A ritual to 'raise his head' must be performed as early as possible, failing which he may even die. And this ritual is certainly one of the most important rituals Limbus perform, and has a striking similarity to the ritual Lohorung Rais perform in order to raise the *saya* or 'ancestral substance'; it is 'one of the links between an individual and the world of the ancestors'. It is also seen as 'the principle of warlike strength' (see Hardman 1981: 162–5).

No wonder then the first chapter of Philippe Sagant's *The Dozing Shaman: The Limbus of Eastern Nepal* is enitled 'With Head Held High'. Actually, as

Sagant rightly points out (Sagant 1996: 15), the word used by Limbus is 'soul' and not 'head', but since they believe the soul rests in the head, the above expression is not inappropriate. Like the Tamangs, Rais, and other Himalayan communities, Limbus have a concept of multiple souls resting in different parts of the body, particularly the head, both the right and left shoulders, the armpits, the chest, etc. (ibid.: 18–19). They also have a concept of *Phung-sam* or 'flower-soul', which is the 'vegetal other', called 'vegetal double' by Sagant (ibid.: 20–1). Whereas the various body souls are considered as seats of power/energy, the flower-soul is considered as the seat of life and indicates the present as well as future of the person it represents. Marriage auguries are made on the basis of flower-souls and divinations about future are often based on it. In the *tongsing* ceremony, which may last from one to three nights, and which is performed by every lineage group once every three years to renew the life force of its members, the entire ritual is focused on such soul. The Limbus believe that if the flower or the 'vegetal other' of a person who is sick, which is kept in a vase (usually made of brass or zinc), withers away while the shaman is performing the ritual, the person it represents will not survive either.

Sagant (ibid.: 19–20) also points out that blood is an extremely important medium for 'raising the head' as it also is for cleansing an offence. Blood is the revitalizer of the soul as well as the body. And whether the raising of someone's head will be done with the blood of chicken, goat, pig, or buffalo depends on the economic and political status of the person whose 'head' has 'fallen'. How hard he beats his chest or thumps his feet as a part of the 'raising of the head' also depends on the status of the person concerned.

The Limbus also have the practice of planting a banana tree on the grave of a stillborn child and cutting the tree before it starts flowering in order to symbolically relieve the mother from the torments caused by the soul of such a child (ibid.: 20).

Conclusion

The previous sections show how the social world of Limbus is imperceptibly blended with the physical and the spiritual worlds in which they live. In fact, there are several concepts like soul, god, deity, spirit, ghost, energy, and power that are used interchangeably and no clear boundaries are ever drawn between them. The spiritual beings can be both benevolent and malevolent depending on whether they are happy or not. But identification of the sex of the supernatural being responsible for someone's ailment is important to ensure that the animals or birds to be sacrificed are from the right sex (for a male deity will need to be propitiated with a male bird or animal only). The supernatural beings are believed to be visible to a Phedangma or to some individuals with the knowledge of communicating with them, but will not be visible to ordinary human beings. Furthermore, these supernatural beings need a medium to reveal themselves to the living. Such a medium could be both inanimate, like rocks, utensils, blowing pipe, wind, winnower, or animate like trees, bamboos, birds and animals. Most importantly, supernatural beings are everywhere – inside the house, in the main post, in the hearth, in the threshold, in the granary, or outside the house, in the spring, in the cow shed, in the kitchen garden, etc. But some such spirits are to be offered sacrifices only outside the house and some only inside the house, some only with blood and some without blood, some with the promise of blood and some without such a promise, some above the house and some below it, and so on. Such knowledge of the spiritual world is not confined to Phedangmas; most elderly Limbus know it.

It is obvious from the above that the Limbu conception of their physical world is interlinked with their spiritual and social worlds. Whereas Limbus may lack the scientific knowledge of interdependence between the various elements of physical environment and thereby cause great harm to it they certainly know how such elements are linked with various elements in the spiritual and social worlds. For, in their worldview, they are not just elements of nature but also elements inhabited and nurtured by spiritual beings whose domain very well extends to human beings as well, both while living and after death.

Such an intermeshed worldview is under threat of extinction today. As elsewhere in African and Latin American countries, the sacred specialists have become extremely rare and those whom Limbus think are experts in matters of their religion are even rarer. One needs to travel much further distances today to find a Phedangma and one should indeed consider oneself lucky if he is found at home at the end of a long journey. The Limbus of the eastern Himalayas are concerned about this crisis but there is little they can do to salvage themselves from such a situation. One cannot become a Phedangma the way one can become a doctor or engineer; he has to be born with such a destiny.[10] So, at least, they believe.

If Phedangmas have become rare today, certain plants that are vital to their rituals and dances have vanished completely due to rampant deforestation all over the region. The wild animals and birds have vanished with the forest and so have the thick bamboo groves that might once have sheltered wandering ghosts and spirits. Some Limbus are, however, surprised that there has been no adverse impact on their life even when they have stopped making sacrifices to so many gods and spirits as their forefathers did. No one organizes a three-night long *tongsing* any more, not because no one can afford it, but no living Phedangma is expert enough to conduct it. Several households belonging to the Samba clan have no he-goats reared for sacrifice in the name of their clan-deity called *Dungdunge*.[11] Where have so many of their gods and spirits vanished? How long will their worldviews last? No one has an answer, nor perhaps can there be one. Perhaps one day the Limbus might come together and perform a ritual to 'raise their head' again.

Notes

1 *Lhomentsongsum* is a 'treaty' between three (*sum* in Limbu language) communities: the Lhopas or Bhutias, Membas/Monpas or Lepchas, and Tsongs/Chongs or Limbus. These three communities had taken an oath to live as one family with the Lhopa or Bhutia as the father, Memba or Lepcha as mother, and Tsong or Limbu as son. This was accomplished by the first Tibetan king in Sikkim, Phuntsog Namgyal, to neutralize the imminent resistance to his rule in the newly consolidated kingdom called Sikkim, which is a corrupt version of the Limbu word *Sukhim*, meaning 'new house'. The Bhutias called it Denzong and the Lepchas, Mayel Lyang. The Limbus had fought against the Tibetan ruler prior to the consecration of this 'treaty'. See Sinha (1975) and Subba (1990: 40–5).

2 Tampungma is one of the most important Limbu deities. This deity is particularly important because Limbus believe that the souls of all adults must pass through the forest to reach the 'lake of souls,' the ultimate destination for the souls of the dead. See Sagant (1996: Chapter 13).

3 Despite strong reservations expressed by Ben Campbell, in Chapter 10 in this volume, against what he calls 'ethnically circumscribed worldviews for understanding environmental relations' (p. 187), I would think that engaging in such a worldview is neither dated nor inferior to an approach focused on 'subjectivities' that Campbell advocates as a better alternative. While I agree with him on the abuse of cultural approaches in environmental studies, I opt for a more informed cultural understanding rather than abandoning the cultural approach without giving it a fair chance to prove itself. After all, as Butz and Eyles argue (1997: 8), environment is, in the ultimate analysis, culturally constituted.

4 *Janajāti* refers to ethnic groups like the Rai, Limbu, Yakha, Magar, Gurung, Tamang, Sherpa, Tharu, and Sunuwar. Since 1990, the members of these communities have launched a full-scale movement in Nepal to revive their language, religion and culture, which were suppressed under the statist agenda of promoting the Nepali language, Hindu religion, and Hindu culture in Nepal.

5 Andrew Russell, in Chapter 3 in this volume, discusses the 'Yakha' perception of environment, arguing that one must understand the social and cultural contexts which give substance to their perceptions.

6 *Tongba* refers to the bamboo container in which fermented millet or wheat is poured, allowed to soak with hot water, and sipped with the help of a bamboo pipe with a small slit at the bottom so that when one sips only the alcoholic content comes up. Almost every Limbu, as much as Rais, Yakkhas and Lepchas, keeps several such containers in the kitchen. During marriage and funeral ceremonies, young men fetch such containers in large numbers from the forest so that every adult guest gets one each. For regular use, many Limbu keep *tongba* made of wood with the edges decorated with aluminium braces. A young boy can sip an adult's *tongba* without getting scolded but never gets one for himself. In this sense it symbolically differentiates the adults from the adolescents. There is, however, no particular age at which one begins to get one and an indulging grandmother might allow even a minor to sip her *tongba*.

7 *Khukuri*, often wrongly spelt as *Khukri* or *Kukri*, is a slightly curved knife with the edge inside and the broad end outside. It has various sizes, lengths and shapes. This is considered an extremely handy instrument in forests as well as in war.

8 *Dasaî* is one of the largest Nepali festivals that occurs during October on the eve of *Vijayadaśami* of the larger Hindu society. During the festival the sons and daughters who have been away because of jobs or marriage return to their natal home seek the blessings of their parents and grand-parents. The sons and daughters are also expected to bring their spouses and children to seek the blessings of their parents-in-law or grandparents. It continues for three days during which

period there is a lot of feasting in every house and the young children go around the houses of their senior relatives seeking the blessings of the elderly and enjoying the ceremonial food. The houses are nicely daubed, whitewashed or painted and there is thorough cleaning of both inside and outside of the house. This is not done only if a senior member of the lineage or family has expired during the year.

9 *Tihār* takes place about a fortnight later. This is an equally important festival of Nepalis in which groups of young men and women go from house to house singing songs, narrating mythological stories, and blessing the members of the household. In return they are offered food and drinks. The houses are especially decorated during this festival with paper flowers, real flowers, banana trees, etc. One of the highlights of this festival is worshipping of birds and animals like crow, dog, cow and ox on different days. One of these nights is a new moon night in which they light oil lamps and candles to welcome Laxmi, the Hindu goddess of wealth. On the last day the brothers seek the blessings of their sisters who in turn receive gifts or cash from them. Limbus abstain from celebrating this festival only if there is death of an adult at home or in the lineage group. However, after 1990s many of them have given up celebrating both *Dasaî* and *Tihār* on the ground that they are actually Hindu festivals. For more details on this, see Subba (1999b).

10 Phedangma, or the Limbu priest, has become a rare institution today. The decline of this institution and its total disappearance from some parts of the eastern Himalayas have, as Limbus rightly argue, nothing to do with clientele. Although Limbus have been influenced by Hinduism and many have converted to Christianity, there is no dearth of Limbus who would seek the services of a Phedangma. Limbus believe that the most important reason for this crisis is that one has to be born with signs of becoming a Phedangma, and such children are simply not being born.

11 *Dungdunge* is the reigning deity of the Samba clan, but interestingly it moves through the female line, from mother to daughter. When a Phedangma identifies the cause of a Samba clan member's sickness to be the anger of his/her clan deity, a he-goat is reared without castrating it in the name of this deity and a ritual is performed when the goat grows to some size. The meat of this animal is distributed among the clan members after the ritual is over. Failure to please this deity can lead to serious illness and even death of a clan member.

7 The role of religion in conservation and degradation of forests

Examples from the Kumaun Himalaya

Safia Aggarwal

Introduction

Lynn White Jr's (1967) paper, "The Historical Roots of Our Ecological Crisis" was significant in raising inquiry into how religious conceptions of the world define human relations and use of the natural environment. Since then, the rapidly expanding literature in ecological anthropology, and most notably the fields of spiritual ecology and environmental philosophy, have attempted to provide insights into the role that religious belief systems play in regulating natural resource use (Tucker 1997; Palmer and Finlay 2003; Taylor and Kaplan 2003).

In particular, literature has focused on the protection of geophysical areas such as mountains and hilltops, rivers and streams due to their association with the divine (Fickeler 1962; Bernbaum 1990). This literature has also documented efforts to protect specific species of flora (Mansberger 1987; Sponsel *et al.* 1998) and fauna (Ghildial and Sharma 1985) through association with the sacred. A vast body of literature has also emerged on sacred groves and temple forests, documenting the intentional or inadvertent protection of various sizes of forests due to the link with the supernatural (Chandrakanth and Romm 1991; Chandran and Hughes 2000; Anthwal *et al.* 2006; Arora 2006; Dhaila-Adhikari and Adhikari 2007).

Using examples from the hill regions of Kumaun, this chapter examines how cultural and religious conceptualization of the natural environment influences human–environment interactions and local use of forests. The first section of the chapter explores how specific natural elements acquire an exalted status often due to association with the divine, and the implications this has for the natural environment. This section shows how the concepts of exalted are relative, and discusses the multiplicity of meanings associated with the sacred. The second section of the chapter examines how societies at times *intentionally* use religion or "supernatural sanctions" (based on divine retribution) in regulating natural resource use, albeit with mixed social and environmental results. It illustrates how ideas of the sacred are interpreted differently by different individuals within the village society.

The research was conducted in a region known as Dharamghar (named after a small market town) in the mid-hills of Kumaun, which is one of the two regions and administrative divisions of Uttarakhand, a mountainous state of northern India. The nearly 15 villages that formed the focus of this study are located at between 1500 and 2000 meters altitude in the border regions of the Pithoragarh and Bageshwar districts, approximately 40 kilometers west of the Nepal border and 80 kilometers south of Tibet. The study was conducted between 1998 and 1999. This research was supplemented by a site visit in 2004.

Primary data for this case study were collected through village and household surveys, and through interviews with relevant Revenue and Forest Department officials. Special effort was made to obtain a diversity of perspectives by interviewing men and women of various age groups, castes, and from various economic classes determined by land ownership and sources of cash income, as well as elected village and forest council members regarding changes in natural resource use. Semi-structured interviews were conducted with key informants such as temple priests, and specific individuals (often elderly) at the village level known for their knowledge of religious texts and cultural norms. Other scholarly works on Kumauni socio-cultural traditions formed an important source of information. Participant observation proved critical to both sections of the study since it allowed for the opportunity to compare information obtained via the interviews with actual practice.

Religion and spiritual ecology in rural Kumaun

Most rural Kumaunis in the research site, including the traditionally migrant Indo-Tibetan communities popularly referred to as the Bhotia identify themselves as Hindus. However, religion practiced in rural Kumaun contains elements of both the Vedic Hindu and the local animistic traditions. Within the Hindu tradition, the worship of Shiva and Shakti (female energy) is most prevalent in Kumaun, and numerous spirits and regional deities of the animistic tradition are commonly worshipped in Kumaun, especially in the rural areas.

The regional deities are very closely connected with practical concerns of rural Kumaunis and are related to the surrounding natural elements. For instance, the Kṣetrapāl Devatā or Bhumiyā is the god of the fields, and protects crops from natural calamities, disease, and damage from wild animals. This deity is also responsible for resolving boundary conflicts related to the division of agricultural lands. Other local deities of primary importance in this region include the Mul Narāyan Devatā and his two sons, the Banjyend and Nauling Devatā, who reside in oak forests and water sources respectively. These local deities play an important role in the lives of the village residents.

Sacred status of natural elements in the Kumauni biophysical world

For a variety of reasons, specific aspects of the biophysical world, including landscapes and species of flora and fauna are given a sacred status by rural communities. For instance, rural Kumaunis often refer to the entire hill region of Kumaun as "*dev bhūmī*" (the land of the gods, marked by the presence of the divine). Moreover, as in many other parts of the world, mountains and hill tops in Kumaun are highly regarded, and given an especially exalted status. These are generally thought to be the abode of the gods and belonging to these divine beings. Thus, the highly revered peak of the Nanda Devi is believed to be the abode of the goddess Nanda Devi. In the research area, even the local hill tops are associated with the divine—as places in local mythology where deities showed their presence—and identified by temples built there, as, for instance, the temple of the deity Mul Narāyan located on the local mountain peak known as Shikhar.

Special significance is also accorded to the rivers, and confluences of rivers (*prayāg*) for similar reasons, evidenced in the presence of the temple Sangar located on the tributary of the River Saryu, and the Shiva temple at the confluence of the Gomti and Saryu rivers in the town of Bageshwar. A feeling of reverence is typically associated with such places, as these places are viewed as the abodes of the divine beings.

Similarly, specific species of wild fauna are highly regarded and therefore protected in Kumaun as in others parts of India, due to the direct association with specific mythical characters. For instance, the eagle (*garūdā*) is revered because it is said to be the carrier of the god Vishnu, and so are snakes (*nāgā*s) due to their association with the Hindu god Shiva. Nagas are perhaps the most commonly worshipped fauna in Kumaun as in other parts of India. While people are generally fearful of *nāgā*s, the *nāgā*s are at the same time given an exalted status out of a feeling of reverence, and are therefore worshipped and protected.

Not all species of flora given an exalted status are associated with the divine; some are viewed as simply auspicious (*śubh*). While most people, including the Hindu priests may not know the specific reasons for the sanctity of specific flora or fauna, it is generally assumed that these floral species must have favorable properties because these species were referred to in religious texts, and were favored by ascetics and sages of the past who were especially knowledgeable about flora (particularly their medicinal properties). Other than the association with the religious texts, mythical characters, or sages of the past, ecologically beneficial qualities (*lābhdāyak*), social value, aesthetics (*śobhā*), or symbolic reasons such as longevity dominate local explanations for the exalted status of these floral species.

Thus, the *pipal* (fig) tree and *bargad* (banyan) trees are venerated for good ground water retention capacity and for purifying the air, and generally considered valuable (*lābhdāyak*). These trees are widely used as shelter and shade for outdoor village gatherings, and revered due to their association with Hindu and Buddhist religious texts. According to local priests, other trees such as the deodar (*devdār*—Himalayan cedar) and the *surai* (Himalayas cypress) are venerated in Kumaun primarily for their beauty (*śobhā*). Deodar trees are considered *śubh* in rural Kumaun, but the chir pine (*cīṛ*) trees belonging to the same family are considered a curse locally due to the dry ecological conditions they create. The *surai*, also a coniferous tree of the *Pinaceae* family, is given a more exalted status than the chir pine but less so than the deodar, despite the three tree species producing similar ecological effects. One local explanation for the difference in protective status accorded to the three tree species was attributed to the aesthetic appeal of the deodar trees with their conical crowns. The surai trees are also somewhat conical in shape, but they are considered less aesthetically pleasing than the deodar tree, and therefore less revered.

Many tree species in Kumaun associated with Hindu religious texts are also trees of exceptional medicinal value, as, for instance, the *bel* (wood apple) tree associated with the Vedic god Shiva, the fruit of which is used as medicine. The much-revered *bhojpatra* (birch), is considered particularly effective in casting charms and spells, and is used in treating snakebites. Not just trees, but numerous other flora including crops such as rice, roots such as turmeric, various herbs, climbers and grasses, many of which also have medicinal properties, are given an exalted status in the Hindu tradition as a whole and in the study region of Kumaun.[1]

Some floral species have a higher exalted status than others. In this continuum, floral species that have a higher exalted status (such as *pipal* or banyan) are revered in themselves and are protected regardless of their physical location. Floral species that are relatively less revered are not necessarily protected unless associated with an exalted natural or built area. These less revered plants are, nevertheless, commonly planted in places such as temple groves. In doing so, these specific plants or trees (and not the species) move up the continuum among the highly revered floral species and are also protected. For instance, a surai tree planted by a child in the temple complex of a local deity in Simgari village was protected by the villagers due to association with the temple. This tree grew and remains today the most prominent tree in the entire temple complex. Table 7.1 provides a summary of the various meanings of the sacred discussed above.

Dev van ***and temple forests***

Dev van (forests of the divine) or temple forests in the Hindu or Kumauni context ascribe to yet another conception of the sacred. These groves are not just associated with the divine as in the case of many sites and flora, nor are they designated or dedicated to the divine, instead they are taken to simply *belong* to the divine. Often when temples were built in remote forests, boundaries were not created to separate the sacred elements of nature. As villages expanded, forests around the temples came under cultivation or became degraded from overuse. In some cases, in the research site, with the initiative of specific individuals, walls were built around some of these temples incorporating a portion of the forested area. This occurred in the case of the once remote Murain Devatā temple in village Airadi, as well as the

Table 7.1 Multiple reasons for ascribing a sacred status to elements in the biophysical world

Reason for veneration	*Forms of natural elements*	*Example*
Dedicated to the divine	Forests	*Dev arpit* forests*
Belonging to the divine	Forests (e.g. *dev van*)	Badrivan
Associated with an exalted place	Particular trees not otherwise venerated	A pine tree near a temple
Śhubh or auspicious (associated with sages)	Forests, specific floral species	Fig tree
Śhubh or auspicious (related to mythology)	Forests, specific species of floral/fauna	Snakes
Ecological benefits (*lābhdāyak*)	Forests, floral species	Fig tree
Medicinal properties (*lābhdāyak*)	Floral species	Birch
Aesthetics	Floral species providing *śobhā*	Deodar
Symbolic meaning (e.g. longevity)	Floral species	Banyan tree
Social significance (e.g. for shelter)	Floral species	Banyan tree

Discussed in the second section of this chapter.

Banjyend Nauling Devatā temple in Udiyari village, where small plots of original forests still remain due to the surrounding wall built in more recent history.

Although the notion of *exalted* flora and *dev van* are considerably different from each other, local reasoning on the presence of *dev van* in Kumaun overlapped considerably with reasons for veneration of the specific floral species. The most common explanation provided by local temple priests on the existence of *dev van* was simply aesthetics (*śobhā*). Other reasons included shade and shelter for the deities particularly on the hill tops, and ecological benefits primarily the retention of ground water and therefore protection of water sources.

Despite the shared perceptions on the notion of the exalted, there were variations in local interpretations of the origins of *dev van*. As noted by the Kumauni historian, Shekhar Pathak (pers. comm. 1998), *dev van* were likely to have been designated for conservation purposes. *Dev van* are generally designated above agricultural areas, because of the local belief that in doing so the deities would protect the fields and the crops from all harm. Thus while villagers attribute good crops to the supervision and protection accorded by the deities, a good crop may be ensured because of retention of ground water resulting from the presence of the groves. The head of the village council (*pradhān*) and priest of Bhatti *gaon* confirmed this belief with the common local expression among Kumauni villagers, that "a hundred trees above will produce water below."[2] Many of the utilitarian functions, however, appear to be by-products of *dev van*.

Much of the literature on sacred groves has focused on the establishment of these areas sometime in the distant past. And, indeed, many of the popular *dev van* in Kumaun came into existence centuries ago under the patronage of the Kumauni rulers. These rulers periodically offered land to temples in gifts (*guṇth*) to gain the favor of specific deities. However, in Kumaun, numerous *dev van* have been established in the past twenty years. In some cases, the temples existed already, and specific individuals took the initiative of reforesting areas around them. This is the case with the Banjyend Devatā temple and grove in village Mahroori. In other cases, individuals established temples and *dev van* where none existed before. Interviews with individuals responsible for designating the relatively new groves suggest that *dev van* were meant to enhance the spiritual integrity of these places, a reasoning that overlaps with that of *śobhā* [aesthetics] discussed above. However, in one instance in village Bongar, a new *dev van* was established by a village resident specifically for the protection of a fresh water source (*jal sroat*) in the midst of the cultivated landscape. To maintain the water levels, trees were planted around the water source, and a small stone temple was constructed to ensure that the trees were not harmed by the other residents of the village. Many *dev van* in rural Kumaun do have fresh water sources within them, but it is difficult to know whether these were established for conservation purposes and/or for mythological and aesthetic reasons.

Local mythology

Several scholars have suggested the role of myth in ecology and adaptation to the environment (Reichel-Dolmatoff 1976; Harris 1979), in the encoding and preservation of knowledge (Eliade 1990) and in the maintaining of moral conduct within societies (Malinowski 1926).[3] Analysis of Kumauni as well as other Hindu myths also suggests functional elements of ecology and ethics embedded in them.

The popular story of the mythical brothers Banjyend and Nauling provides a useful illustration from the research site. The two *devatā*s, considered twins by many villagers, are believed to be the sons of Mul Narayan Devatā. According to a local shaman (*dungariyā*) of Simgari village, well known in the region for his extensive knowledge of local and Vedic mythology, the elder son of Mul Narayan Devatā became known as the Nauling *devatā* as a result of his confrontation with a demon in the nearby village of Sangar. In the intense struggle that followed between these mythical figures, Mul Narayan's son gained victory over the demon, upon which an uncarved stone (*liṃga*) associated with Shiva and at times with local deities appeared in the water catchment [*naula*] in Sangar. Since then he is known as the Naul-ling *devatā*, and hence the Nauling Sangar Mandir was built to remember the event.

The temple is now visited annually by villagers in the region. The younger son of Mul Narayan is believed to have been born from the core of an oak [*banj*] tree, and is therefore referred to as Banjyend Devatā. Today temples devoted to these two brothers are typically seen together, built in a single temple complex, and referred to collectively as the Banjyend Nauling temple. The two temples are normally identical, so much so that most villagers are unable to distinguish between them.

Thus, both in the ecological and the mythical realms, the water and the forests (particularly oak) are closely linked, and the *naulas* and oak forests are perceived to coexist. It is possible that the ecological relationships were captured in local myth intentionally with the purpose of instilling these concepts among local residents, but this can only be hypothesized.

Religion and environmental degradation

The exalted status of certain aspects of the biophysical environment may not always result in the protection of natural elements, and indeed may contribute to degradation. For instance, customarily in Kumaun, only dead wood extracted from the *dev van* could be used in ritual ceremonies. Today, however, felling live trees for ceremonial practices is not uncommon. I witnessed this during a ceremony at the Ajandi Bubu Mandir in the Lohathal village where two standing live trees in the vicinity of the temple were felled for a night-long ritual ceremony [*pūjā*].

In some instances, the rising popularity of specific temples, religious mass tourism, and resources used in large ritual ceremonies contribute to the depletion of the once remote *dev van*. The *dev van* of Hokra, en route to Munsiyari in Kumaun, provides a case in point where the enormous expanse of the *dev van* surrounding the temple has been considerably degraded due to the rising popularity of the temple and excessive harvesting of leaves of specific trees for making leaf plates [*pattal*] traditionally used to serve consecrated food offerings [*prasād*]. Many sacred places in Garhwal that are now linked to roads are today facing even greater pressures from religious tourism. These temples, once surrounded by vast tracts of *dev van*, are now expanses of complexes with only a few remaining trees.

Furthermore, in Kumaun an even greater problem of environmental degradation due to association with religion may be located in the changing nature of contemporary religion itself. In Kumaun, as in other parts of rural India, local religion has been changing due to the Hinduization (rising emphasis on Vedic deities) and Brahmanization (rising importance of the priestly caste) of the local society, and hence the increasing emphasis of contemporary religion on the Vedic rather than the animistic traditions. This in turn is resulting in an increasing separation between nature and religion, with greater importance being placed on the material aspects of religious culture in place of the natural and the supernatural. As a result, even the Vedic deities that were traditionally associated with certain natural elements, such as rivers, are being dissociated from the natural elements and increasingly being worshipped as idols.[4]

In addition, despite the significant role of the local deities in people's lives, there is an increasing association of these deities with the Vedic gods and goddesses. As, for instance, in the research area, villagers increasingly associate and identify the Kotgari Devī, the local goddess of justice, as one of the eight forms of Shakti. Not only are local deities increasingly associated with the Vedic deities, but natural elements of worship are being replaced by deities of the Vedic tradition. This association with the Vedic deities can be seen in the shifting representation of the local deities from lithic elements to human forms. Thus, deities of the animistic tradition which are typically represented in the form of uncarved sacred stones [*liṃga*], believed to have self-originated (*swayambhu*) or, having been brought from a river bed, to be pure (*pavitra*) are being replaced by idols in human forms, representing the Vedic gods and goddesses.

Combined with these changes in local religion and religious practice is the change in contemporary notions of aesthetics that places greater importance on the material rather than on natural elements. In rural Kumaun, many *dev van* are being replaced by large temples and temple complexes. In the Dharamghar region, numerous older temples were replaced by newer ones during the early 1980s through the initiative of a well-known local priest. During this period, the priest, with local support, had a large temple constructed in place of the ancient temple of the regional Mul Narayan *devatā* and surrounding forest area at Shikhar. Similarly, in Sangar, in 1991, much of the oak grove was cleared to accommodate the large temple of the Nauling *devatā*. In Kumaun, emphasis on the temple and temple structures has also facilitated grazing in the *dev van*. With greater significance given to temples rather than the groves, the temples remain unexploited, while the groves experience extensive grazing pressures. Chandran and Hughes (2000) note similar changes in religious practices and the move away from exalted natural landscapes to built landscapes in other parts of India.[5]

Intentional use of the sacred in conservation

This chapter has so far focused primarily on unintentional consequences of religion on human–environment relations. Religion and notions of the sacred in relation to the biophysical world are also in many cases used *intentionally* as a means of conserving natural elements, such as for ecological restoration. Such intentional use of religion in natural resource conservation is often attempted with the help of (or by) priests and other influential individuals who ascribe religious significance to activities to ensure support by the masses. According to several respondents in the field site, this strategy was more prevalent in the past, when people had greater faith in the efficacy of religious and supernatural beliefs than people do today. Nevertheless, religion continues to be used today in rural areas specifically for problems related to environmental degradation, health or hygiene that appear to lack secular solutions (see Aggarwal 2001, for various regional examples).

Given that most decision-making processes are inherently political and often controlled by the elite, intentional use of religion is worth a closer look. Who uses religion to create, or enforce existing rules of natural resource use? Who decides the rules of natural resource use under the aegis of religion? What are the social and environmental consequences of such intentional association with religion? The next section provides a closer look at one case where religious faith and fear of the supernatural are being used in an effort to address environmental concerns.

Intentional use of religion and the supernatural in Pañcāyat *forest conservation*

In Kumaun, many *van pañcāyat* [forest councils] have been facing problems of excessive forest degradation, particularly in oak forests. During the 1990s, lopping of oak leaves for fodder had been increasing, resulting in poor regeneration of oak trees. In the Dharamghar region, excessive degradation of *pañcāyat* forests, and the inability of the *van pañcāyat* to enforce rules of forest use led to the use of supernatural sanctions and transfer of surveillance of *pañcāyat* forests to a local deity [*dev arpan*] by a number of the forest councils.[6]

In 1992, one village decided to place its forests under the care of the local deity popularly known as the Kotgari *devi.* On a designated day members of the *van pañcāyat* were invited to participate in a circumambulation of the forest boundaries, and the case was presented to the local temple of the Kotgari *devi.* At the temple, a ritual ceremony [*pūjā*] was performed under the directions of the temple priest, and the deity was appealed to, to oversee the forest. No longer was the forest watched over by forest guards, and no longer were violations of forest rules questioned. It was believed that since the appeal, the deity presided over the forest, and that divine retributions would strike any user daring to transgress the rules of forest use.

By 1998, approximately 30 *van pañcāyats* had placed approximately 2200 hectares of *pañcāyat* forest area under divine protection. Five of the 30 *van pañcāyats* formed the focus of this study. Placing forests under divine protection took place primarily in the oak zone due to excessive degradation from lopping of trees for fodder. Each *van pañcāyat* placed under the protection of the deity its entire *pañcāyat* forest for a period of five years, except those *pañcāyats* that had pine forests under their jurisdiction. Villagers collect fuelwood in pine forests, but with sufficient dry wood available, lopping of pine trees is rare, and few *van pañcāyats* in the pine zone experienced the degradation that was seen in oak forests.

Justice system of the local deities

As noted earlier, faith in the mythic realm and the power of deities form strong elements in local thinking in rural Kumaun. Although the younger generation may have less faith in the supernatural, misfortune in their own lives will often strengthen their belief in, and fear of, the deities.

In the Dharamghar region, the Kotgari *devi* is the deity of justice, and the temple of this deity is locally referred to as the *nyālaya* [court], where individuals who have experienced injustice may seek divine ruling. Belief in the efficacy of the deity's justice system brings to this temple numerous local cases of conflict. Although in much of rural Kumaun, criminal cases and land conflicts are resolved through the regional revenue officer [*patvārī*], villagers often appeal to the divine when the revenue officer is unable to resolve the case. Rural, and at times urban Kumauni, often prefer to approach the deities for justice rather than the appropriate human authority due to the stronger belief in the divine powers. The poor in particular will appeal to the deities to avoid the problem of submitting a case and incurring the expenses of government courts. Appealing to the deity may be especially effective where the offender may confess misconduct in fear of divine retribution.

As with other deities, the justice system of the local deities is based on divine retributions. Violations of rules associated with the divine may result in various types of mishaps and unanticipated misfortune for the guilty individual and his or her household. These misfortunes may relate to health, such as unexplained sickness or untimely death of a member of the household or of cattle, but may also take the form of crop failure, financial loss, or unexplained disappearance of belongings. Examples of the powers of Kotgari Devi and of other deities were well known in the region. Indeed, several instances of divine retribution that took place during the course of the research illustrate local faith in the supernatural, the power of the local deities, and the process through which the supernatural system of justice worked. In this link with the supernatural, the *dungariyā* (shamans) play an important role in identifying cases of divine retribution.

The following case of divine retribution narrated by a woman of Lohathal village is particularly insightful. I once visited the informant while she was preparing for a visit to Dudhil *gāon*, the village of her in-laws and she invited me to accompany her. During the trip she related a recent experience of divine retribution on her husband's household. Her husband, a migrant in Mumbai, had broken their child marriage, and had remarried in the city. The informant, aged 12 at the time of the incident, suffered considerable ill treatment from her in-laws' family and was finally sent back to her natal village. In her anguish she made a plea to the *jal devī* [local water goddess]. The informant was taken by surprise, however, when without having formally appealed to the *devī* for justice, divine retribution took place twenty years later.

Several members of the husband's household became afflicted with an incurable illness. After several unsuccessful visits to the hospital, a shaman was consulted by the husband's family. The shaman "read the rice" provided to him by the afflicted family by placing grains of uncooked rice in the palms of his hands, and moving them while uttering incantations and names of the local deities. The pattern of movement of the grains was interpreted as a message from the deities. The shaman informed the in-laws of the plea made by the daughter-in-law to which the *devī* had responded. They were asked to perform a *pūjā* [ritual ceremony] to appease the *devī*. After much pleading by the in-laws' family, the informant agreed to participate in a *pūjā* at the temple, and the guilty members of the household acknowledged their mistakes. The informant, however, refused her husband's offer to bring her to the city. The case was finally resolved when the informant accepted the husband's offer to build her a house in the in-laws' village. By the time of this visit, the informant had also begun receiving monthly compensation from the husband. Meanwhile, two members of the in-laws' household passed away while the others recovered. The visit to the informant's in-laws' village confirmed this story. The consequences and future were still being discussed during my visit. As the informant explained, she felt reluctant about believing stories of divine retribution, thinking them to be superstitious, but with such an incident taking place in one's own life it is impossible to ignore it. During our return, the informant ended the story with the reasoning that "after all, the Himalaya is *dev bhūmi* (land of the gods), and that gods live in the forests, the hills and the Himalayas, and that therefore it is no surprise that justice prevails here." The only thing, she added, was that it may take quite some time before justice is effected. This incident illustrates local faith in the divine system of justice, and the closely connected worlds of the human and the supernatural.

Local perspectives of dev arpit pañcāyat *forests*

Although local perceptions of *pañcāyat* forests under divine protection (*dev arpit* forests) varied, most residents noted the functional nature of giving the *pañcāyat* forests to the deity, viewing it as analogous to the customary means of resolving conflicts in the absence of secular means of enforcing forest rules. This explains the reason for handing over entire *pañcāyat* forest rather than only sections of it. Asking the deity to supervise only parts of the village forest (as I proposed during the interviews) had no meaning for most villagers. For most people, *dev arpan* represented primarily a seeking of divine assistance in the management of the problem. Thus, in Chaunala village where management of pine forests posed as much of a concern as of oak forests, pine forests were also placed under the deity's protection, despite the fact that pine forests are generally considered a curse.

Despite the similarity between placing forests under divine protection and appealing to the deity in cases of land conflict, the *dev arpit pañcāyat* forests were unique in their temporal dimension. Most residents viewed placing forests under divine protection as a management strategy, where management of forests remained under human rather than divine authority. This is reflected in the fact that unlike in cases of land conflicts, *pañcāyat* forests were placed under divine protection for a limited period of time, rather than until a resolution was reached. The head of the forest committee [*sarpanch*] of one village, expounding on the management function of *dev arpan*, noted that *dev arpit* implied constant vigil by the deity over the forests, providing a more effective mechanism of surveillance than the system of forest guards. Placing forests under divine protection reduced the risk of disrupting social relations and eliminated the cost of regulating, monitoring and enforcing forest rules.

Dev arpit pañcāyat forests may appear similar to *dev van* [forests of the divine] described in the previous section, but the two remain conceptually and practically very different for the residents. For most people, even new *dev van* that continue to be designated in their villages remain considerably different from the *dev arpit pañcāyat* forests.

As villagers explained, *dev van* are almost always associated with temples, and are typically left around temples for *śobhā*, that is, for aesthetic reasons. These trees or *dev van* need not be exalted in themselves, but are revered in so far that they are associated with a temple and hence presence of the supernatural. As noted earlier, some of the new *dev van* in the region have in fact arisen out of ecological concerns, however, in those cases only the individuals seeking divine protection may be aware of the reasons for it.[7] Moreover, *dev van* are typically not entrusted to the deity (*dev arpit*) but simply "belong" to the deity. Generally, *dev van* are exalted but divine protection is not sought for these areas. Since they are not placed under divine care they also cannot be formally removed from divine protection. *Dev arpit pañcāyat* forests, on the other hand, despite having been placed under divine care, remain *use* forests and entrusted to the divine for a limited time. *Dev arpit pañcāyat* forests may have numerous temples devoted to the various deities lying within them, however, these forests are not *associated* with any of the temples. Ownership of the *pañcāyat* forests continues to rest with residents. It is only protection of these areas that had been accorded to the divine.

While many phenomenologists and scholars of religion have described feelings of veneration towards particular places or biophysical features (Isaac 1967; Tuan 1978), seeking divine intervention in the mid-hills of Kumaun may be better understood as inscribing a sense of fear of the divine [*ghāt* in Kumauni or *śrāp* in Hindi]. *Dev arpit pañcāyat* forests are associated with divine protection, but not with the *presence* of the divine. Also, fear associated with these *dev arpit* forests is not fear of the *landscape* itself (Tuan 1979), but of breaking of taboos resulting from the association with the deity. For the residents, what were earlier rules of forest use had now become taboos in the sense of a "contained potency" described by Shore (1989), and the danger and dread associated with the violation of the taboos. Interestingly, in *dev van*, even plucking a leaf from a tree might be considered potentially dangerous. In *dev arpit pañcāyat* forests, the inadvertent plucking of a leaf from a tree has no significance since the intent of this act does not violate the rules associated with divine supervision. In other words, danger in the case of *dev arpit pañcāyat* forests lies specifically in the *intent* of violation of rules of forest use.

Typically no formal rules of resource use exist in *dev van*. However, implicit rules of use associated with these groves coincide with those of the *dev arpit pañcāyat* forests. In both cases, only dry resources may be extracted, and green resources may be extracted only for use in ceremonies performed in a temple complex contained within the specific forest. Nevertheless, rules regarding felling trees for ritual use remain unspecified for both cases of forests.

Social and environmental implications of dev arpit pañcāyat *forests*

The impacts of *dev arpan* were complex due to inter- and intra-village variations in ecological impacts and household adaptations (see Aggarwal 2001), nevertheless in practice, placing of *pañcāyat* forests under divine care resulted in two relatively minor changes in rules of forest use. Under the forest guard system, villagers were permitted to collect oak leaves for fodder on specific days during the winter months. In placing forests under divine care, forests were no longer opened for fodder collection. In addition, the forest guard system allowed for the removal of understory, especially thorny shrubs, for use as fuelwood. With protection of forests given entirely to the deity, the removal of green understory was no longer deemed acceptable. As under the forest guard system, however, villagers could continue to collect fuelwood by lopping only *dry* branches of trees and shrubs, or by collecting fuelwood from the forest floor.

These minor changes in rules resulted in significant shifts in local patterns of natural resource use. Placing forests under divine protection dramatically reduced incidences of encroachment in *pañcāyat* forests when compared with the forest guard system. In particular, incidences of lopping in the *dev arpit* forests declined as residents began relying on less preferred forms of fodder, such as grasses, as the primary livestock feed. Some households chose to reduce the size of their cattle holdings due to the difficulties of collecting fodder from the *dev arpit* forests. The environmental impact on fuelwood collection were notably different, however. With few fuelwood alternatives in the region, and restrictions regarding the use of only the dry shrubs, residents found it difficult to meet their daily fuelwood needs from the *dev arpit pañcāyat* forests. Many households therefore began extracting fuelwood from the adjacent *pañcāyat* forests, government forests, or private forests in the region. Thus, *dev arpit pañcāyat* forests were conserved, but collateral impacts of fuelwood collection were significant as the problem shifted to adjacent forest areas.

Placing forests under divine care resulted in unexpected difficulties for most households. Alternative places of resource collection were located at considerable distance from the villages and were dominated by pine trees not suitable for fodder collection. Poorer households with smaller landholdings and cash incomes, however, are typically more *dependent* on forest resources than the wealthier households. Therefore divine protection placed differential pressure on the various classes of village society. For the wealthier households, supernatural sanctions posed inconvenience in the daily collection of forest resources. Nevertheless, with larger landholdings, these households increasingly resorted to collection of green leaf fodder from the private forest plots, and private trees on cultivated lands. Wealthier households also own larger plots of cultivated lands, and therefore have greater amounts of grass and crop residues available to them as livestock feed. Additionally, with greater access to cash incomes, these households were also able to access the relatively expensive market alternatives to fuelwood such as kerosene. For the poorer households, sanctions meant hardships in meeting basic needs without the security of the alternatives. With little cash income, and no private forests and trees, these households had the greatest difficulty meeting their fodder needs.

Using religion in forest management: decision-makers, forest users, and transgressors

The use of religion as a mechanism for asserting control over areas and resources within them have been discussed by other authors (see Mawdsley, Chapter 8 in this volume; Gezon 1999). Indeed, political relations in the form of local power and wealth differentials have implications in the management of any resource.

In Kumaun, decisions regarding management of *pañcāyat* forests lie in the hands of the *sarpanch*, and to a certain extent the *pradhān*, both typically representing the elite within the village social structure. As noted above, the elite households have greater access to forest resource alternatives than do the poorer households. There remains, therefore, a gap between the decision-makers (typically the village elite), and the poorer households who have a greater dependency on forest resources than do the elite. There remains also a gap between decision-makers and forest users in terms of gender. With many of the men now engaged in the cash economy, women remain the primary collectors of forest resources in many hill villages, while men (often the elderly) continue to be the primary decision-makers. Indeed, in the case study, while men were responsible for making the decision on seeking divine intervention (particularly in the first village), the impact of the decision was primarily on the women in terms of increasing women's time and labor in the collection of forest resources. During my interviews, while women did not explicitly state concerns regarding the placing of the forests under divine care, they did acknowledge the difficulties of collecting resources from neighboring forests. In addition, while forest degradation has a significant impact on resource collection for most people, regardless of economic status or gender, nearly all interviewees appeared to be greatly concerned about depletion of forest resources. Nevertheless, placing forests under divine care is not a solution that many would have opted for if given the choice.

A careful look at the first village to place its forests under divine protection suggests that this decision came from the head of the village council and the *van pañcāyat* committee. Entrusting *pañcāyat* forests to the deity in this village led to a rise in encroachment in adjacent *pañcāyat* forests. Since most villages in the region had been experiencing difficulties with the forest guard system, many villages had little choice but to seek divine assistance in the management of their own *pañcāyat* forests. Thus, while in the first village, seeking divine assistance was a result of decision-making of the elite men, in the subsequent initiatives the decision to seek divine assistance was more inclusive of the poorer households and of women. In some villages, the women and the poorer households themselves initiated placing their *pañcāyat* forests under divine protection due to the concerns of resource depletion.

Transgressions of the *dev arpit pañcāyat* forests perhaps best reflect the politics of placing forests under divine surveillance. Violations of the rules of use of the *dev arpit* forests suggest that a wide range of households and at times organized groups engaged in transgressions. Some violations of the rules were carried out by the poorer households, often without the knowledge of other villagers. In Mahroori village, violations to the rules of *dev arpit pañcāyat* forests had been taking place on a routine basis. Here the poorer Shilpkar (the Scheduled or low caste) households of the Saknyura *toak* (sub-division) engaged in organized transgressions. My knowledge of this was derived from an accidental disclaimer made by a resident of Saknyura who disclosed that their entire *toak* was engaged in such transgressions in Mahroori *dev arpit* forests. The *pañcāyat* forest of Saknyura village was small, and had over the years reached such extreme levels of degradation that the forest had not even been considered for divine oversight. For the past twenty years, Saknyura villagers had been collecting all forest resources from the civil forests (in theory, government forests co-managed with village communities). Because Mahroori forests had been entrusted to the local deity and the forest guard removed, Saknyura transgressions had shifted to Mahroori *pañcāyat* forest due to its proximity. Rather than engaging in household level transgressions, specific individuals from Saknyura village collected resources from Mahroori forests distributing them to other households in the *toak*.

Transgressions were not limited to the poorer households, however. In some villages such as Airadi, the number of households opposing the decision to seek divine assistance was quite large and represented the relatively wealthier households. Upon the forest council's placing of the *pañcāyat* forest under divine surveillance, this group of households rebelled by performing the sacrilegious act of removing the village temple bell and placing it at the doorstep of the *sarpanch*'s house. This in effect was a declaration to the decision-makers that divine protection would not prevent them from breaching the forest rules. Some open dissent to sanctification and consequent transgression also took place in Lohathal village, where the dissenting groups (also the elite) was relatively smaller than in Airadi village. Rather than taking a strong political stance as in Airadi, this group merely rationalized its transgressions by claiming that they had not taken part in the placing of their *pañcāyat* forest under divine care and were therefore not liable to the rules emerging from it.

While these stories of group transgressions were not difficult to come by, identifying cases of individual transgressions was more difficult given the secrecy associated with such acts. Some residents may have known the identity of the transgressors, but the local *dungariyās* were particularly knowledgeable of transgressions in cases where divine retributions had taken place. Since *dungariyās* are responsible for diagnosing divine retribution, villagers consult with these spiritual functionaries regarding all cases of unexpected misfortune.

Stories of divine retribution due to violation of sanctified forest rules were few but were well known in the region. One of the better-known instances was from Mahroori village where the sudden death of a bullock belonging to a wealthy Rajput (a high caste) household was identified as divine retribution for the lopping of oak leaves from the sanctified forest. In general, interviews with the *dungariyās* confirmed what most residents noted, that households on the wealthier end of the socio-economic spectrum were more often violating forest rules than were the poorer households. The dominant role of the wealthier households in transgressions suggests that many transgressions took place because of the inconvenience of accessing alternative areas rather than out of material necessity posed by the lack of alternatives. It also suggests that wealthier households were more open to taking risk than the risk-averse poor. As the *sarpanch* of Mahroori village noted, wealthier households felt that they had little to lose in risking divine retribution. According to the *sarpanch*, if divine retribution resulted in the loss of a goat, rather than being disturbed by it, these households would be pleased to consume the extra meat. Although the logic poses a rather simplistic view of the level of harm that could result from divine retributions, economic insecurity and fear of divine retribution may very well have a positive correlation.

The transgressions of the *dev arpit pañcāyat* forests appeared to increase over the years. This may be attributed to the fact that some of the difficulties of accessing the *dev arpit pañcāyat* forests emerged later as the forests regenerated. Restrictions on lopping shrubs led to regeneration of forest understory in these forests. This in turn made access to dry resources from the *dev arpit* forests difficult, possibly prompting greater rule violations. In other cases, the transgressors set an example for the other households that divine retribution need not be feared.

Regardless, unlike rule violations under the forest guard system that snowballed into a competition, where each user attempted to capture resources before others did, transgressions to rules of *dev arpit* forests remained relatively controlled. A general understanding was that unlike penalties for breaking the forest guard system, divine retributions were not negotiable. Moreover, the belief that retribution could transpire within a ten-generation span in the future, kept transgressions in check.

Current state of **dev arpit pañcāyat** *forests*

In 2004, of the five villages that formed the focus of this study, only one village (Simgari) continued to use the deity as a means of managing the forests. This village used the experience to modify the concept to meet local needs. After five years of divine surveillance, the village again sought divine assistance. However, it was decided that only the northern section of the forest would be placed under divine care for the first three years. Upon completion of the three years, the southern section of the forest would be placed under divine protection for two years, thereby dividing the five-year period between two sections of the forest. Thus, in this village, the use of the strategy had attained an increasingly managerial characteristic, and as a result provided viable livelihood alternatives to the village residents. This village was also relatively successful in limiting the collateral impact on adjacent forests.

Upon completion of the five years, two of the five villages joined the Joint Forest Management program of the Forest Department. The remaining two villages decided to return to the forest guard system of guarding the *pañcāyat* forests. Interviews in these two villages suggested that instances of encroachment in these *pañcāyat* forests remained limited as people feared that rise in encroachment might instigate the decision-makers to again seek divine assistance.

Conclusion

This chapter illustrated how, in rural Kumaun as in many other societies, religious beliefs and notions of the sacred influence human relations with the natural environment. It illustrated the multiple meanings of exalted elements in the biophysical world due to association with religion, and the relativeness embedded in the notion of the exalted in religion, local mythology, and in actual practice in people's relations with the biophysical world. The second section of the chapter showed how religion and faith in divinity continue to play an important role in dealing with problems that lack ordinary solutions in rural Kumaun. Conservation of *pañcāyat* forests using religious means presents a case in point.

The chapter showed also how association with religion may be helpful in conservation, but cannot be *assumed* to lead to positive outcomes. This is because religions may not always prescribe a conservation ethic. Where a conservation ethic *is* embedded in the religious philosophy, religious rituals may take precedence over the conservation ethic. This is particularly the case where there is dissociation of nature and religion, and a move away from the animistic to the material or transcendental aspects of religion, as seen in contemporary rural Kumaun. In other cases, economic and livelihood concerns, rather than religion, may play a greater role in determining human actions towards the environment, resulting in a gap between the religious conservation ethic and human environment interactions. In ascribing an exalted status to an element in nature, local intent rather than religious belief is critical in determining the ecological outcome of these revered places, particularly in the long run. This is especially the case where mass tourism or emphasis on built structures take precedence over conservation and maintenance of the sacred within the biophysical world.

Intentional use of supernatural sanctions for conservation of particular areas may be effective where faith in religious beliefs and powers of the supernatural remain strong, and where sanctions do not impinge on local livelihoods. Where viable and equivalent livelihood alternatives exist, people are likely to make use of alternative resources rather than risk breaking supernatural sanctions. However, when these sanctions impinge on local livelihoods without providing alternative resources, they are likely to fall short of their promise. As illustrated in this case study, using divine assistance in the regulation of forest resource use may temporarily limit encroachment, but benefits may be limited to the short run or degradation may shift to adjacent areas. Moreover, when religious faith is intentionally used in conservation, supernatural sanctions are more likely to be effective on the poorer than the wealthier households because of the greater risk-taking behavior of the wealthy.

The elite may in some cases use religious institutions to enhance their political control or to grant preferential access to particular resources. In other cases, as presented in this study, such sanctions may not be *motivated* by political control, however, these sanctions may fail to serve the interest of the entire village, given the differential nature of hardships faced by residents from closure of the forests. Religious or supernatural sanctions like other forms of controls, and decision-making in general, are likely to be more effective when underlying causes of the problem are addressed. Collateral impacts and excessive pressure on adjacent forests in this case reflect the limitations of inadequately addressing the root problem of degradation itself. Entrusting forests to the care of deities may be effective in the long run rather than temporarily limiting encroachment, if it includes planning for alternative resources and institutional adaptation to changing socio-economic circumstances, allowing greater inclusion of those collecting the resource or with greater dependence on it, in the decision-making process.

This chapter, therefore, emphasizes the need to critically examine the use of religion and the supernatural in relation to natural resource use and conservation concerns. Understanding local perspectives of the sacred, and re-establishing the link between nature and religion may be helpful in the protection of the biophysical.

Notes

1 Extensive literature exists on plants that are considered sacred in India and in other parts of the world. See Bargali *et al.* (1995) and Pande *et al.* (1990) for a discussion of sacred flora of Kumaun, and Pandey (1989) and Hajra (1987) for sacred flora of the Himalayas.
2 While recent literature has shown that contrary to popular belief, forest regeneration leads to reduced water runoff (Calder 2000), the phenomena is different in forests that are frequently covered with clouds or mist (as is the case in Dharamghar). In these forests, extra inputs of water to the local ecosystem are allowed by "cloud interception," that is, canopy interception of clouds. In such cases, water deposition from clouds and mist may raise net precipitation and water runoff (Bruijnzeel and Hamilton 2000).
3 According to Eliade (1963), myths are stories about issues that directly concern those who make them. New myths are created to address concerns of contemporary times, while old myths get reinterpreted based on contemporary needs. Such reinterpretation of myth is discussed by Prime (1992) in the case of the Yamuna river in northern India where god Krishna as a child gained victory over a mythic snake that had poisoned the river and all life forms within it. Today, this myth has been reinterpreted by environmental activists who equate that poison with the pollutants emerging from the factories and sewers in and around the city of Delhi.
4 Haberman (2000) discusses a similar phenomenon taking place in the case of the River Yamuna where increasing pollution of the river is leading to the dissociation of the river from the river goddess. The spiritual is seen, and projected by the priests, increasingly as transcendental rather than as linked with the natural.
5 While this trend appears to be more common today, similar conversion from exalted natural to built landscapes was not unknown in the past. For instance, Saraswati (1998) describes the case of Bodh Gaya, where the forest in which Buddha performed meditation and attained enlightenment was replaced by the city of Gaya, and only one Bodhi tree remains, said to be from Buddha's times. Similarly, Eck (1982a) discusses the conversion of revered forests of *Anandavana* (Forest of Bliss) to the holy city of Banaras.
6 See Aggarwal (2001) for a discussion of supernatural sanctions in relation to other types of sanctions (e.g. legal and social) used in regulating natural resource use.
7 Olson (1997) makes the distinction between *aitu* and *tapui* in the Samoan spirit-world, based on the intent of placing biophysical elements under divine supervision. *Aitu* implies obstacle or barrier due to religious prohibitions imposed on a resource for the purpose of maintaining favorable spiritual relations with the gods, such as in the case of sacred groves. Ecological conservation in these cases is only a by-product. *Tapui*, on the other hand, is seeking divine supervision with the purpose of imposing restrictions on the use of the resource with the intent of conservation. Although no such distinction appears to exist in rural Kumaun, the two distinct purposes of seeking divine supervision do seem to exist in the local understanding of these places.

8 The abuse of religion and ecology

The Vishva Hindu Parishad and Tehri Dam

Emma Mawdsley

Introduction

The relationship between religion and the environment is now a well-established subject of academic discussion (White 1967; *Worldviews* passim; Gottlieb 1996; American Academy of Arts and Sciences 2001), and has attracted formal responses from all of the major world religions (WWF 1986). Hinduism,[1] the focus of this chapter, has been the subject of particular attention – recent analyses include Prime (1992), Nelson (1998a), Baviskar (1999), James (1999), Chapple and Tucker (2000), and Narayanan (2001). Within these and other studies, three broad and overlapping analytical approaches can be identified: (1) the exegesis of religious and historical texts to uncover metaphysical and philosophical teachings on the environment, as well as indications of past attitudes and behaviours; (2) ethnographic explorations of particular social and cultural practices which relate to the environment; and (3) sociological analyses of movements and organisations that actively mobilise religious values and beliefs in pursuit of various environmental goals. Although much of this work is broadly positive about the role religious belief can and sometimes does play in encouraging greater ecological sensitivity, some commentators rightfully express cautions and caveats. These include the limitations of textual analysis in elucidating actual values and behaviours, past and present (Pedersen 1995; Nelson 2000; Patton 2000); the evidence that some elements of Hindu belief and practice can be at best neglectful and at worst positively damaging to the environment (Nelson 1998b; Alley 2000, 2002); and questions over whether religion or culture is an appropriate or sufficient frame through which to tackle the massive environmental problems confronting modern India (Nanda 2002; Tomalin 2002). This chapter is concerned with another caution – of a different order and with a different set of implications – the abuse of religion and ecology in India through the mobilisation of 'green' issues in ways that are intended to promote a chauvinist Hindu nationalist agenda. The context for this argument is the growth of Hindu nationalism in the Indian polity and society, and the increasing violence and discrimination against religious minorities and others in India.

The chapter starts with a quick overview of recent analyses of Hinduism and ecology, focusing on the problems mentioned above. It then turns to the rise of the Hindu Right in India, and outlines the way in which values and traditions of accommodation, plurality and syncretism are being suppressed by the increasing strength of a chauvinist, masculinist and violent religious nationalism. At the heart of the chapter is a case study of the way in which the Vishva Hindu Parishad (VHP; World Hindu Council) has sought to communalise the issue of Tehri Dam, setting it within a framework of Muslim/outsider threats to the Hindu nation, while neglecting the social and environmental issues that concern the dam-affected people of the region, and which motivate the wider network of anti-dam activists. There are examples of organisations that genuinely seek to promote social and environmental justice in India through the mobilisation of various religious teachings and values (Roy 1993; Sullivan 1998; Haberman 2000). However, this chapter argues that in the context of an increasingly aggressive Hindu nationalism, environmental issues and movements are by no means immune from the dangerous and divisive religious politics that are being played out in contemporary India.

The arguments presented here draw on a wider research project of the past two to three years, which has focused on the environmental values and behaviours of India's middle classes (Mawdsley 2004). It also builds on earlier work in and around the area of Tehri Dam, from 1992–96, while researching the regional movement for a new federal State in this area. Environmental issues were an important element of the study (Mawdsley 1998), and during this time I lived for several months in a village in the dam's proposed submergence zone. As well as talking to men and women in this and other villages, I interviewed several people associated with the anti-dam movement, and various supporters of the dam, locally and in the wider region.[2] The work of Mukul Sharma (2001a, 2002) has been especially valuable in providing detail on the VHP and Tehri Dam, and I would like to formally record my debt to his analysis.

Hinduism and ecology: cautions and caveats

Approaches to the study of Hinduism and ecology tend to fall into three categories, although often with very considerable overlaps. The first focuses on the hermeneutic analysis of various religio-mythological and historical texts, such as the Vedas, the Upanishads, the Epics and many more, in order to elucidate various philosophical arguments and precepts concerning the environment, as well as indications of past behaviours (e.g. Dwivedi 1997; Mumme 1998; Sharma 1998; Dwivedi and Tiwari 1999; Rukmani 2000; Lutgendorf 2000). Second are the more ethnographically based analyses, which focus on: (a) particular *practices*, such as the maintenance of sacred groves (Gadgil and Chandran 1992; Sinha 1995; Freeman 1999); (b) on specific *issues*, such as waste disposal (Chaplin 1997; Agarwal 2000); and (c) on the environmental associations and outcomes of the material *expressions* of Hinduism, such as caste (Chaplin 1997; Korom 1998) or rituals (Khanna 2000; Nagarajan 2000). A third set of writings, ethnographic and socio-political, examines the ways in which particular organisations and movements have mobilised various Hindu idioms and beliefs in the pursuit of particular environmental goals. Examples include Swadhyaya in Gujarat (Roy 1993; Gupta 1999), and various groups encouraging the reforestation of the Braj region, a place special to devotees of Krishna (Haberman 1994, 2000; Sullivan 1998). Environment-livelihood movements, such as the Chipko struggles (Mawdsley 1998; Rangan 2000b), and the Narmada Bachao Andolan (Baviskar 1995), to name but two of the better known, also fall into this category. Of this wide, and often very rich literature, the best emphasise the importance of situating beliefs, practices and movements within local and regional and historical contexts; the sheer diversity of Hindu theological traditions, beliefs and practices; the intersections with issues of gender, class, caste and community; and their hybridisation with other religions and practices in and from the sub-continent, including Islam, Christianity, Buddhism, Jainism, *sarna*,[3] and so on (Nelson 1998b; Apffel-Marglin and Parajuli 2000; Jackson and Chattopadhyay 2001; Narayanan 2001).

Many of these analyses question and complicate the popular assumption that Hinduism, like other 'eastern religions', necessarily encourages values and behaviours that are environmentally positive – an argument that is often set against a critical reading of Western philosophical and religious traditions in relation to the environment (White 1967). First are the various limitations circumscribing the 'ecological interpretation' of historical and religious texts. Most scriptures, commentaries, poems, legal treatises and so on, reflect the views of their authors, who were usually elite, high caste, educated men. Although they provide a valuable source of material, interpretations of past environmental values or behaviours cannot be inferred simply by reference to Brahminic (or later Mughal) texts and treatises. They are most unlikely to have reflected the full array of behaviours among elite, high caste men themselves, and are certainly inadequate to examine those of subaltern groups (including women, lower castes and *ādivāsis*), across the different regions, folk cultures and language groups of the subcontinent. In any case, a range of contradictory points can be sustained by selective reference to these texts – they are rich and diverse, spanning history, regional culture and opinion. Lutgendorf (2000), for example, examines tales of forest destruction in the Ramayana and Mahabharata, some of which are told with a certain morbid relish in the scenes of violence (see also Kelkar and Nathan 1991; Grove *et al.* 1998); while Jha (2002) finds plenty of evidence of meat and even beef eating within early and later texts of Hinduism, Buddhism and Jainism (see also Zimmermann 1987). A third problem with the exegesis of texts is that while all world religions make claims about the foundational nature of human–environment interactions, and include certain prescriptions and regulations that relate to the non-human, they do not directly or systematically address 'modern' environmental challenges (Tomalin 2002). The many forms and traditions of Hinduism pre-date the emergence and growth of the qualitatively new ecological problems that now confront the world, including nuclear threats, global warming, chemical poisoning, and technologies that allow extraction, processing and pollution on a previously unimagined scale.

None of these observations invalidate the search for philosophical positions, examples or teachings that encourage environmentally positive behaviours, such as respect for plant and animal life, and ecological humility. However, such an enterprise must be seen as a conscious, pragmatic construction in the light of contemporary problems, rather than the recovery of some 'ancient ecological wisdom'. This critique is particularly relevant to many 'new traditionalists', as Sinha, Greenberg and Gururani (1997) describe them, who tend to construct an idealised, romantic vision of past social and environmental harmony, and build upon it a vision for an alternative future. As well as their historical inaccuracy and social conservatism (such arguments often ignore the evidence of human-induced ancient and pre-colonial environmental change, and neglect or even condone historical inequalities of gender, class and caste), these 'new traditionalists' have difficulty in presenting realistic agendas for engaging with modern India – its industrialisation, urbanisation, and the widespread growth of consumer desires among all social classes (Varma 1998; Gupta 2000; Osella and Osella 2000; Mawdsley 2004). A 'return to the past' is simply not an attractive or viable option for most of India's urban and wealthier population, while its poorer masses are very frequently doing all they can to escape the hardships of 'traditional' life. This is not a defence of current social, economic or political structures, but a recognition that the struggle for social and environmental justice must engage with these realities.

A further set of problems arises from the fact that Hindu beliefs can encourage what might be thought of as positively 'anti-environmental' attitudes. For example, Alley (1998, 2000, 2002), has explored in detail the different cognitive frameworks that operate – and clash – around the pollution of the Ganges. As a central site and symbol of holiness for Hindus (and others) across India, the Ganges is worshipped by many as the all-cleansing Mother, capable of absorbing and forgiving any sin brought to her. But as Alley points out, although the river is venerated, this does not necessarily translate into concern about its gross material pollution, in the non-ritual sense, by sewage and a whole variety of hazardous wastes. The transcendental purity of the river cannot be sullied. This is one reason why scientists, local activists, foreign donors and (some) concerned bureaucrats have consistently failed to win public interest in or support for their efforts to clean up the 'immanent' or 'profane' river.[4] On a smaller scale, many authors have noted the consonance between 'Hindu' attitudes to bodily purity and pollution, and the practice of dumping garbage on the street. One's own space/body is kept scrupulously clean/pure, while rubbish/pollution is expelled, to places/people who are not one's concern (Chaplin 1997; Agarwal 2000). Nelson (1998b) also questions the celebrated 'holism' of Hinduism, which supposedly stands in opposition to the transcendental dualism (mind/matter; human/nature) of the Semitic religions, and therefore – supporters argue – promotes a humble recognition of the divine in all things and respect for all creation. Nelson argues that one of Hinduism's principal theological traditions, Advaita Vedanta, is indeed founded on a dualism, in which the physical and conscious world is to be renounced in the quest for *moksha*, or liberation from birth and re-birth. The world of consciousness and matter, or *maya* (usually glossed as 'illusion'), is worthless and to be despised – hardly a propitious foundation for environmental concern (see also Sherma 1998, on the gendering of *maya*).

These problems have been widely discussed, and urge caution in thinking about the relationship between Hinduism and 'pro-environmental' values, attitudes or behaviours. To reiterate, for many commentators, this does not mean that the potentially positive relationship between Hinduism and ecology is de-legitimised, but that promoting more 'positive' environmental ideas and actions through the idioms and understandings of Hinduism(s), must be understood as a situated, actively constructed, and complicated enterprise.[5] However, a rather different and very serious problem arises when we consider the way in which the Vishva Hindu Parishad, a key actor in the Hindu Right, has sought to communalise the issue of Tehri Dam, drawing environmental issues into the fundamentalist rhetoric of Hindu nationalism. Before exploring this in more detail, the actors and ideology of *Hindutva* will be introduced briefly.

Hindu nationalism

Since the early 1990s, the growth and consolidation of Hindu nationalism have transformed the social and political landscapes of India. Hindu nationalist organisations date back to the 1920s,[6] and despite a low profile at times, they have played a significant part in India's colonial and post-Independence history (Jaffrelot 1996; Ludden 1996; Hansen 1999; Zavos 2000; Bhatt 2001). Together the various elements of the Hindu Right are known as the Sangh Parivar, a 'family' of organisations and parties. The main constituents are the Bharatiya Janata Party (BJP); the Rashtriya Swayamsevak Sangh (RSS), a 'cultural' and voluntary service organisation founded in 1925 which promotes the paramilitary-style training of men and women in drill and martial arts; and the Vishva Hindu Parishad (VHP), or World Hindu Council, which was set up in 1964 as an attempt to bring the diverse elements of Hindu belief and practice under one ecumenical umbrella (Katju 2003). The 'youth wing' of the VHP, the Bajrang Dal, is composed mainly of disaffected young men who can be called upon to intimidate and threaten opponents. In addition, there are many other smaller political and non-political affiliates. Like all families, the Sangh is divided on certain issues. This can be seen as a weakness – senior BJP figures appear sometimes to be embarrassed by the violent and illegal actions of their partners, and sometimes they are distinctly under pressure from the RSS and the VHP to adopt more extreme policies and agendas. However, these differences can also be seen as a strength. In the complex game of India's sub-continental federal politics, the BJP is able to affect compromises and make pragmatic partnerships with ideologically unlikely allies in order to extend its appeal in different regions and to a wider range of social groups (Basu 1996; Datta 1999). The BJP is by no means always successful, and a number of authors point to the difficulties it faces in transcending its biases towards north Indian, Brahminic Hinduism and social elitism (Chiriyankandath 1998; Manor 1998). But even as it (usually) acts and speaks in more 'moderate' and accommodating ways, it supports and takes political advantage from the politics of hate espoused and acted out by the RSS, the VHP and others.

The core ideology of the Sangh is that of *Hindutva*. In direct translation, this broadly means the 'essence of Hindu-ness', but it has come to refer to the more narrow agenda of 'Hindu nationalism'. Essentially, *Hindutva* represents a 'blood and soil' vision of the sacred land of Hindustan for the Hindus. Although, inevitably, there are a range of positions within the Sangh, the dominant view is that Buddhists, Jains, Sikhs and *ādivāsis*, whose origins are in India, can be absorbed within the Hindu fold. Christians and Muslims, however, who supposedly have no organic connection to the fatherland, must be expelled, or can stay only in complete subjugation to Hindus and in compliance with the ideologies and agendas of Hindutva (Jha 1999). This is a Hinduism that is assertively militant and masculine in tone. The early Hindu nationalists, such as Veer Sarvarkar, opposed the 'effeminacy' and non-violence of Gandhian nationalism, and this anxiety over and preoccupation with virility continues today (Alter 1992; Hansen 1996). During the Gujarat riots of early 2002, for example, in which around 2000 people (mainly Muslims) were killed, and tens of thousands were displaced, sexual violence was widespread and clearly symbolic of the tropes of masculinist nationalism. The aggressive tone and nature of contemporary Hindu nationalism are also evident in the celebrations of the 'Hindu [nuclear] bomb', which was tested in 1998. The aim was in part to intimidate Pakistan, but the event was also strongly oriented towards winning domestic acclaim by asserting a new aggrandising status for militant Hinduism (Corbridge 1999; Das 2003). Many ideologues of Hindutva also seek to promote a more homogenised and doctrinally united set of beliefs and practices, downplaying and in some cases actively suppressing the plural, accommodationist and syncretic nature of much 'folk' Hinduism, while promoting more masculinised festivals, symbols and figures as India-wide expressions of Hinduism.

Analysts of the Hindu Right have examined the diverse and often apparently contradictory attitudes to economic liberalisation (McKean 1996; Hansen 1998; Jenkins 1999; Harriss-White 2003); its gender and gendered ideologies (Sarkar 2002; Banerjee 2003); and the ways in which the Sangh has sought to expand its social and political support amongst lower castes and *ādivāsis*, and in the south of India (Basu 1996; Jaffrelot 2003). Some of the key debates about Hindu nationalism concern the nature of 'secularism' in India (Madan 1987; Bhargava 2003); the extent to which the Sangh Parivar might be considered fascist (Bhatt 1999); and the place of the BJP in the changing political economy of India (Corbridge and Harriss 2000). Of particular concern to many is the violence and discrimination against India's religious minorities that has accompanied the rise of the Sangh Parivar (Baber 2000; Misra 2000; Melanchthon 2002). Sumit Sarkar (2002) suggests that although the BJP is subject to political setbacks and reversals, and there is profound resistance to the ideologies and agendas of Hindu nationalism in India, the Sangh is managing to persistently and insidiously change the terms of public and political debates in India (see also Hansen 1999).[7] As we shall see below, this extends to certain environmental issues and movements, and we turn now to the case of Tehri Dam.

The VHP and the Tehri Dam struggle

Tehri Dam has been under construction since 1978. It is located just below the confluence of the Bhilangana and the Bhagirathi rivers (which go on to feed the Ganges) in the Himalayan State of Uttarakhand. The dam has attracted widespread criticism on environmental, social, economic and safety grounds. At 260.5 metres, it will be one of the tallest dams in the world, but is being built in a V-shaped valley in the geologically young and seismically active Himalaya mountains. Some engineers have expressed their deep concern about the possibility of catastrophic failure (e.g. Brune 1993). If it were to give way, the immediate and downstream consequences would be immense – whole towns could be flooded in a matter of hours, potentially drowning and displacing tens of thousands. As well as these specific features, Tehri Dam is subject to the familiar and depressing litany of problems associated with large development projects, and especially dams. Its benefits – power, irrigation and drinking water – will mainly accrue to richer groups in the Gangetic plains, including industrialists, larger farmers, and urban dwellers, while its costs – displacement from land, livelihoods and communities, without adequate rehabilitation or compensation, as well as ongoing environmental damage – will be borne mainly by the poor, women and rural people of this marginal mountain region. The substantial illegal flows, in the form of bribery for construction contracts and rehabilitation management, are mainly being harvested by social and political elites. The costs of the project have inflated over the years, while estimates of the profits have eroded downwards (International Rivers Network 2002). Other critical facts and figures have also been subject to repeated revision by the authorities. For example, the Progress Report on the Rehabilitation of Oustees stated that, as well as the town of Tehri itself, 22 villages would be totally submerged, and 59 partially affected. However, the THDC's 'Rim-Line Survey', undertaken between November 1990 and July 1991 indicated that Tehri and 21 villages would be submerged, while 75 villages would be partially affected. At the same time, some THDC officials were apparently privately estimating that the reservoir would totally submerge 25 villages (Singh 1992). Estimates from opposition groups tend to be higher. At a *dharnā* (sit-in) and hunger strike in Tehri town in February 1994, I was told that 192 villages would be affected to some degree. Sunderlal Bahuguna has drawn attention to the difference between official wavering over the figures for displacement and submergence, and the certitude and precision with which electricity and water supply metrics were stated (Interview 22 April 1994). Over the past 20 years there have been periodic halts in construction, as various court cases and investigations have been conducted into aspects of the dam, but despite repeated concerns and negative conclusions, work on the site re-commenced and the dam is now functioning.

Some people in the region argue that the dam has brought much-needed jobs, investment and development. There are few opportunities for formal employment in this mountainous State, and the limitations of subsistence agriculture mean that migration remittances are an essential part of the hill economy. However, ever since the inception of the dam, local and regional groups have protested against it, supported by wider networks of social and environmental activists (Kishwar 1992, 1995; Friends of Chipko n.d.). The fact that the dam is located in the same area as the famous Chipko movement (Shiva 1989; Mawdsley 1998; Rangan 2000b) has also helped in giving it a reasonably high profile among activists and scholars in the West. A key figure in the anti-Tehri Dam movement is Sunderlal Bahuguna, a veteran campaigner for social and environmental justice, who, as we shall see, has become an increasingly controversial figure. The anti-dam movement has waxed and waned over the years, and there have been splits over whether to fight against the building of the dam, or whether, more pragmatically, to lobby for a better compensation and rehabilitation package. During 1994, I lived for some months in a village that was due to be flooded by the dam. Men and women there were anxious and depressed, and there was considerable tension within the village over their threatened future, the difficulties of staying united, deciding what strategy to take, and choosing who should negotiate with the authorities. The stopping and starting of the project, and the lack of transparency over rehabilitation arrangements made a bad situation much worse.

In the mid-1990s, a new actor came on the scene, as the Vishva Hindu Parishad started to pay increasing attention to the Tehri Dam issue. For many, the Ganges is India's holiest river, visited and worshipped along its length, particularly at certain sacred confluences and sites (Eck 1982b). The VHP claims that its principal contention is that the dam will compromise the goddess – checking her flow will change her fundamental nature and limit her celebrated self-purificatory abilities.[8] The VHP also argue that if the dam fails, the resulting disaster would wash away and submerge many of Hinduism's holiest sites and, among the tens of thousands of others, drown thousands of the *sants* and *sādhus* who congregate along the river (Sharma 2001a, 2002). At first glance, the arguments of the VHP against the dam might be considered a good example of Callicott's (1994) call for environmental issues to be framed in culturally resonant ways, drawing as they do on powerful religious and cultural beliefs and idioms. This is an appealing argument, both in terms of its potential effectiveness – efforts at improvement and/or encouraging local participation might be more successful because they draw in ordinary people in ways that have meaning for them; and in terms of its postcolonial politics, given that it constitutes an alternative to the technocentric, reductionist and economically centred philosophy of mainstream sustainable development. But in this case there are a number of problematic features that raise difficult questions about religion and ecology: the use of communal and anti-Muslim rhetoric; the neglect of local peoples' concerns; and the apparently short-term nature of the VHP's interest.

Mukul Sharma (2001a, 2002) notes that although the holiness of the Ganges was invoked in earlier protests, this was set within a more secular discourse on the beauty of nature, and was secondary to the social and environmental issues of displacement and damage (see, for example, Kishwar 1992, 1995). However the metaphors and idioms that are being deployed by the VHP are deliberately oriented towards an inflammatory anti-Muslim rhetoric. Sharma (2002) states:

> in the later part of the movement especially, anti-Tehri dam politics has persistently and centrally been constructed through a conservative Hindu imagery, often in partnership with Hindutva politics . . . The ecological reasoning is blurred and goes beyond logic, eliciting Hindu support, patriotism and xenophobia.

Sharma quotes a *sādhvi* (a woman renouncer) addressing a crowd in Delhi in 2000, setting the issue in terms of a Hindu–Muslim struggle:

> The Tehri Dam is being constructed to imprison the Ganga forever. This is an organised conspiracy to demolish our religion and culture. The way we had to demolish the Babri mosque [at Ayodhya] at our own risk, we have to get ready now for the demolition of the Tehri Dam.

The comparison with the Babri *masjid* (mosque) is highly revealing. After an aggressive campaign, militant Hindus illegally demolished a sixteenth-century mosque in the north Indian town of Ayodhya in 1992, claiming that Muslim invaders had built over a former temple marking the birthplace of the god-king Ram. In the widespread rioting that followed, over 2000 people were killed. The events at Ayodhya mark a deeply symbolic moment in the changing political landscape of India, and it still acts as a referent for the political debates and manoeuvrings of the Hindu Right, as well as ongoing communal violence. Situating the Tehri Dam struggle in relation to the Babri Masjid is making a clear threat to Muslims, symbolically locating them as foreigners whose presence and influence must be purged from the nation. It is also a signal of intent to the more accommodationist elements of the BJP. For example, a newspaper report quotes the Chair of the Ramjanmabhoomi Trust (a Hindu nationalist organisation seeking the building of a temple on the site of the old mosque), who also made a specific connection between the dam and the Babri Masjid, while revelling in the VHP's influence in government. He said:

> LK Advani is the first Home Minister in India to have come to Hardwar to wash our feet. He will follow us . . . [but] . . . if the government goes ahead with the Tehri dam project, I will do to it what I did to the Masjid on the December 6th 1992.
>
> (*Indian Express* 1999)

According to Sharma (2001a), VHP leaders have claimed that the construction of Tehri Dam is a foreign plot to weaken and attack the Hindu nation. As well as checking the flow of the Ganges, they suggest that it is a prime target for a terrorist attack. Shivaji Rao, a professor of geology who writes on the Tehri Dam issue for the VHP website, has said elsewhere:

> [A]ny bombing by the agents of enemy countries or militants or terrorists, the dam being made of earth, sand and gravel, will easily collapse, killing millions of Indians and destroying the holy temples and sacred towns and cities . . . In view of the latest terrorist attack on the World Trade Center and the Pentagon buildings in the USA a similar attack by terrorists, militants or enemies of Tehri Dam should be anticipated because it will be an inexpensive bid for them to easily destroy the life and culture of the whole of *Āryavartha* [Hindu culture].[9]

The very real dangers of the dam are being conflated with a threat to 'Hindu culture' and the 'Hindu nation'. The Hindu Right thrives on raising anxiety and concern – about Muslim population growth, Christian and Islamic religious conversion, sexual predation – and so on, and then manipulates these fears into social and electoral power bases. By situating Tehri Dam within these tropes, they seek to twist it to their agenda.

Rajalakshmi (2001) reports that the response from local people has been mixed. Some feel that having the VHP on their side will help their cause, given its proximity to power and political leverage. Thus, Sunderlal Bahuguna has been increasingly closely associated with the VHP, courting their involvement in his ongoing battle against the dam. Bahuguna has always been controversial – he has been criticised for being rather media-savvy and, more seriously, for presenting the Chipko protests as the defence of 'traditional' livelihoods and ecologies, especially by women, while not doing enough to foreground the economic and village industry aspects of the movement (Rangan 2000b). He has also been criticised for his anti-dam protests by some in Tehri town and the valley, who feel that his intransigent opposition has undermined demands for a better rehabilitation package.[10] Nonetheless, Bahuguna has many supporters, and has fought all his life for the wider goals of social and environmental justice in the hill region. Inspired by the ideas and methods of Gandhi, he has adopted non-violent tactics in his battle with the powerful. But now, for many secular and Gandhian activists, he is disturbingly compromised by his ties with the VHP.

Rajalaksmi argues that many people whose land and homes are threatened by submergence resent the side-lining of the displacement issues that confront them. For the VHP, the social, economic and environmental issues confronting the villagers and townsfolk are secondary – if perceptible at all – to the main theme of the 'threat to Hindutva'. For example, Sharma (2001a) quotes Ashok Singal, a very senior figure in the VHP, dismissing other concerns while making an 'organic' connection between *river–Hindu–nation*: 'I am not here to talk about the seismic condition of the dam site or about the cost-benefit analysis of the project. I am talking about Gangatva. Gangatva is Hindutva. Hindutva is Rashtratva.'

Singhal has repeatedly iterated this 'connection' between the river, Hindu religion and the civilisation. For example: 'For, if the Ganga is lost, its purity is lost, the Teerthas on the Ganga will be lost and the great Hindu culture will be lost' (quoted in *Times of India* 2000). This observation resonates with a wider criticism of the Hindu Right, which is their limited concern with poverty or the structural bases of inequality, such as caste, class and gender. For example, in a discussion of foreign aid, 'DN' critiques the militant aspirations of Hindu nationalism, and suggests that: 'Politically the poor can be kept in thrall by the vicarious thrills of Indian expansionism [the author refers to the "Hindu bomb"] while economically they remain neglected' (2003: 2353).

Finally, there is clear evidence that the VHP's interventions are primarily politically motivated, with their involvement rising and falling in tune with the political logics and timetables of the State and national elections rather than in relation to dam developments. For example, although Kelly Alley's research is primarily concerned with the Ganges as it runs through the plains of northern India, and specifically the city of Benares, she does briefly turn to the 'upstream' issue of Tehri Dam. Alley argues that the high profile politician-holy men who took up the cause of the dam in the 1998 elections have demonstrated little ongoing commitment to the issue, or to the environmental questions that underlie it. Her analysis is that Tehri is merely an expedient topic, which has been deployed in election campaigns, notably by the VHP's Swami Chinmayananda, and then dropped. This seems to confirm the peripheral nature of 'environmental' issues for the VHP,[11] as opposed to their longer-term interest in power, and in promoting the Hindutva agenda. Sharma (2001a) also mentions Swami Chinmayananda, who was at one point a BJP central government minister, quoting him on his refreshingly frank reasons for opposing Tehri Dam: 'The flow of the Ganges ensures enormous cash flow in our ashrams from Hardwar to Rishikesh. If something goes wrong to this flow here [at Tehri], the flow of money will dry up.'

Writing in early 2002, Rajalakshmi noted that the protestors agitating against the closure of the dam tunnels in December 2001 were not this time joined by the VHP. He suggests that the VHP, 'failed to gain any significant support from the people of Tehri for its 'Gangatva' campaign . . . and it was a matter of time before it would withdraw from the agitation'. This is despite the fact that earlier in the year Ashok Singhal had threatened to go on a fast against the closure of the tunnels. The VHP called off their protests when the Central Government set up another Committee to reappraise the dam in terms of the seismic threat and the purity of the river (its remit did not include the rehabilitation issue, which was not something on which the VHP was lobbying). Predictably, the Committee was only a placatory and brief stalling mechanism, and although the closure of the tunnels has since gone ahead, the VHP appears not to have returned to the issue, disappointed by their failure to whip up more anti-Muslim sentiment. The impact that the dam might have on the divine river Ganges is one that genuinely alarms devout Hindus across India and beyond. The inconsistency of the VHP's involvement suggests that even the faithful have been let down by an organisation more concerned with pragmatic power politics, and to demonise Muslims and others, than to protect the Ganges.

Conclusion

In concluding, it is useful to make a couple of comparative observations with Alley's analysis of the pollution of the Ganges. Alley's starting question may appear to contradict the line of discussion in this chapter. She asks, 'Why have religion and the environment remained *separate* domains given the struggles to control other sacred spaces in India, and the increasing conflation of religion and politics'? The answer, she suggests, lies in the fact that Muslims cannot easily be constructed as the enemy in terms of the environmental pollution of the Ganges because Hindu industrialists act in the same way as Muslim industrialists, and because many Muslims do evidently respect and bathe in the river. For this reason, the VHP and Hindu nationalists in Benares and other plains cities have not developed a strong 'environmental' concern. In the mid-1990s, for example, an environmental NGO persuaded the VHP, which was about to launch a series of India-wide processions, to adopt a call for the Ganges to be saved from its increasing material pollution. But during the processions the issue received little attention, mostly because, Alley argues, the VHP's real mandate was to mobilise anti-Muslim feelings. She concludes that concern for the condition of the Ganges:

> took a back seat to the interest in reinforcing the idea of the Hindu community's Muslim enemy and using this identification with a common hatred to fuel the yatra. The latter interest never appropriated the former, leaving concern for the Ganga a nonissue.
>
> (Alley 2000: 222)

Although this chapter has examined the way in which 'the green and the saffron' (to use Sharma's phrase) *have* come together, in one sense it supports Alley's conclusions – the VHP's concern with Tehri Dam is not environmental, humanitarian, social or economic. Rather, their interest and involvement are driven by the communalising potential the dam issue has when rendered through the idioms and narrative frameworks of Hindutva. Whereas the pollution of the Ganges, for various reasons, has not been claimed as a Muslim plot to weaken Hindus and therefore the Hindu nation, Tehri Dam has been constructed, absurdly but nevertheless dangerously, as just such a threat.

Having made this point, then, we must ask whether the VHP's involvement in Tehri Dam is a cautionary but exceptional coming together of the Hindu Right and the 'environment'. There are, after all, a number of reasons why the Hindu Right may find it hard to colonise 'environmental' issues. First, most 'environmental' issues in India are in fact livelihood issues – contests over access to and control of resources such as land, forests and water (Gadgil and Guha 1995; Guha and Martinez-Alier 1997). The real divides are between rich and poor; multinational, national and local; urban and rural; high caste and low caste; and men and women, which are not easily amenable to mobilisations around Hindutva. Second, as Baviskar (1995) and Basu and Silliman (2000) demonstrate, some environmental struggles actively oppose the agendas and ideologies of Hindutva. They report on *ādivāsi* groups fighting the damming of the Narmada, who are also contesting Hindu nationalist discourses that claim they are 'backward Hindus'. The 'modernisation' and 'Hindu-isation' of *ādivāsis* are viewed by some supporters of the Narmada project as a positive outcome of their displacement (Basu and Silliman 2000). In the defence of their livelihoods, land and homes, some *ādivāsis* are also articulating a cultural politics of resistance to the over-powering embrace of Hindutva. Third, as Alley concludes from her analysis of the Ganges, the impact of various *sant*-politicians has been confined to mobilising public sentiments and electoral support around the 'transcendent' river. Their views have not permeated the scientific-bureaucratic domains of those who oversee the management of the river and dam, including the Ministry of Environment and Forests, and bureaucrats and departments associated with the Ganga Action Plan and Tehri Dam Hydel Corporation. Thus not only is the VHP's 'environmental' interest short-lived, but it is confined to social discourse, producing little in the way of dialogue with or impact on the administrative arm of government.

However, there are indications that this might be too sanguine an outlook given the Sangh Parivar's systematic and insidious efforts to construct a communalised social and political landscape, even if this is contested and only partially realised. In a different paper I have argued that post-development theory provides a very problematic basis from which to contest the discursive claims of Hindu ideologues, including in relation to narratives of environmental history (Mawdsley 2006; see also Sinha *et al.* 1997; Nanda 2002, 2003). In this context, it is vital to recognise that environmental discourses and politics have the potential to be enfolded within the dangerous ideologies and agendas of Hindutva. Although not yet widespread, there are other examples of 'environmental' struggles being caught up within a communal frame. Sharma (2001b), for example, suggests that some battles over coastal fishing resources are being drawn along Christian/Hindu lines in parts of South India. Referring to the finding that many soft drinks in India had unacceptably high levels of pesticide residues, the 'Hindu Unity' website asks the question 'Soft drinks pesticides or Muslim sabotage?'[12] As India's chronic environmental problems worsen (Agarwal *et al.* 1999), so too will struggles between poorer groups. Given the Sangh's groundwork and encouragement, it is possible that in some parts of India, minority and marginalized groups – especially Muslims, Christians and *ādivāsis* – will be increasingly targeted in these 'environmental' and livelihood contests over resources.

These possibilities raise problematic questions about the mobilisation of religious beliefs in pursuit of environmental goals. For example, one reason widely offered for the failures of the Ganga Action Plan (GAP) is that the scientific rationalist principles and pronouncements call forth little interest from most ordinary people. If the GAP is to make any headway, it is suggested, then it must engage with these alternative conceptualisations, and try to work through the idioms and beliefs of Hinduism. But, as this chapter suggests, great caution needs to be exercised. To date, the mechanisms and institutions of India's formal environmental governance – as laid out in environmental legislation and administered by the bureaucracy – have not been strongly affected by the social and religious mobilisations of the Hindu Right. In the present context, this is just as well. In other spheres and places, where elements of the administration are more 'Hindu-ised', the results have been devastating. The police were accused of assisting rapists, looters and murderers, in the 2002 Gujarat riots, by refusing to intervene to assist Muslims, and even by positively assisting the mobs.[13]

Environmental politics are not yet a major site of intervention from the Hindu Right, but the VHP's involvement in Tehri Dam is a worrying demonstration of what to expect if and when they are.

Acknowledgements

I would like to thank the ESRC (Environment and Human Behaviour Programme) for funding the research that, a little tangentially, stimulated this chapter. I am grateful to Glyn Williams, as co-organiser, and all of the participants who came to the workshop on 'India's middle classes and the environment' in August 2003, who contributed to some of the ideas presented here. My thanks also to Arjun Guneratne and the participants of the workshop on culture and nature in the Himalaya (August 2004); and to Clare Palmer and the two anonymous referees who commented on this chapter. Finally, I am grateful to the editors of *Worldviews* who allowed me to reproduce my paper (Mawdsley 2005) here, with minor adjustments.

Notes

1 Writing about 'Hinduism' in such a brief chapter is inevitably simplifying. Not only is it unmatched in its diversity of beliefs and practices across India and beyond, but religion, culture and social form inextricably permeate each other (for example, through the dynamic and varied notions and practices associated with caste).

2 However, I should make it clear that at this time I did not interview these people with the issue of Hindu nationalism in mind, but with regard to the struggles in the region over governance, development and the environment.

3 This is the name given to the 'animist' religion(s) associated with some *ādivāsi* ('tribal') groups. Often these are syncretic with local expressions of Hinduism, Christianity and other belief systems.

4 Other reasons for project failure include the sheer scale of the problem, corruption, and turf battles between different central, state and municipal bodies.

5 Although for a dissenting, original and very stimulating voice, see Meera Nanda (2002, 2003). Nanda argues forcefully that if India's women, low castes and oppressed are to achieve liberation, India must experience its Enlightenment, and banish religion from the public sphere. As long as religious precepts enforce the idea of 'natural' inequality in terms of caste, gender and ethnicity, these groups will always be held back. I discuss Nanda's ideas in more detail in Mawdsley (2006).

6 There were also earlier precursors, such as the Arya Samaj, formed in 1875. For an excellent analysis of the relationship between colonialism and the politics of Hindu identity and religion, see Joshi (2001).

7 Although other parties, notably the Indian National Congress during and following the leadership of Indira Gandhi, must bear a significant proportion of the blame for the increasingly visible intrusion of religion into politics. Despite secularist protestations, Congress has repeatedly played the 'communal' game in various political struggles (Kohli 1990).

8 There is a precedent for such anxieties, when in the early nineteenth century, the British planned to dam the Ganges at Hardwar. For an account of the protests raised, and the British response, see Alley (2002). Interestingly, Alley argues that the religious idioms deployed in fact masked an anti-colonial agenda.

9 See online at: www.gitam.org/science/envstud/tehri/tehridam.html. This was accessed in July 2004, but no longer appears available.

10 This observation is based on various interviews and conversations in Tehri town and four nearby villages over 1993–95.

11 Although this is a familiar pattern in Indian politics in general, and by no means an exclusively environmental observation. As ordinary people are all too aware, politicians come to them when they want votes, with promises and blandishments that rarely materialise after the elections.

12 See online at: www.hinduunity.org/ (accessed July 2004, but now no longer archived).

13 See the 'Report of Concerned Citizens'. Online. Available at: www.sabrang.com (accessed March 2009).

9 Restoration and revival

Remembering the Bagmati Civilization

Anne M. Rademacher

Introduction: a deteriorating riverscape

During the 1990s, unprecedented urban growth brought dramatic change to Kathmandu's physical environment.[1] The Bagmati and Bishnumati rivers, which converge in the heart of Nepal's capital, were among the urban features characterized in official and popular discourses as increasingly degraded.[2] Worsening river conditions motivated a variety of state- and internationally sponsored restoration initiatives that, in turn, drew steady criticism. Some critics charged that official restoration activities were consistently insensitive to, or abusive of, the rivers' cultural significance.[3] Others pointed to the ways that river restoration projects threatened the security of tens of thousands of landless migrants settled on exposed riverbed lands in the riparian zone. Few would dispute that the rivers were degraded, or that degradation had both scientific and cultural facets. Yet precisely why river quality continued to deteriorate, and exactly how to reverse it, were points of considerable contestation.

River degradation intensified during a period of extreme political volatility in Nepal. Over the 1990s, Nepalis witnessed changes ranging from the reinstatement of democracy in the wake of a "People's Movement" (Raeper and Hoftun 1992; Whelpton 2005: Ch. 7) to the violent intensification of a revolutionary "People's War."[4] The contingent nature of the state during this period had particular implications for the capital city, and, in turn, for contests over its territorial character and meaning. Urban river debates reflected and incorporated the political unrest, underlining crucial connections between environmental territories and the reproduction of state power.[5]

The data in this chapter are drawn from a study of the natural and cultural re-territorialization of the Bagmati and Bishnumati rivers between 1996 and 2003. Placing the ideas of Gupta and Ferguson (1992) in the context of urban ecology, the study emphasizes the constructed and contested character of environmental place-making on the Bagmati and Bishnumati, and the range of meanings and identities that were attached to the rivers as ecological places. I explore how the languages and practices of ecology defined specific entitlements to the riverscape, and marked struggles over the conceptual and practical boundaries of culture and nature. Invocations of identity, competing versions of history, and the reproduction of specific ritual practices all figured prominently in river debates, demonstrating the centrality of cultural languages of power to understanding how ecological ideas are enacted in social practice.[6]

This chapter focuses on a specific historical narrative prominent in river debates, the idea of a "Bagmati Civilization (*bāgmati sabhyatā*)." Drawing on textual and participant observation data, I describe the synthesis and circulation of this narrative, noting how it relates the Bagmati River's ecological vitality to particular cultural and political practices. I highlight how claims about the past—and an impassioned appeal to memory—marked aspirations for a particular spatial and territorial order. As such, the Bagmati Civilization produced a moral geography[7] that was at once socially exclusive and politically defiant.

At the center of the Bagmati Civilization narrative was an assertion that the demise of Kathmandu's rivers was integrally connected to a collective forgetting of socio-natural history. Only through 'remembering' that history could the natural integrity of the rivers be reclaimed; a more desirable, ordered polity was expected to follow.

Insofar as the Bagmati Civilization narrative circulated in a crowded field of alternative histories and claims to river territory, we might understand it as one among many instances of dynamic, and perhaps even democratic, 'memory work' (e.g. Samuel 1994). But the Bagmati Civilization idea acquired its rhetorical primacy and power in a particularly volatile historical moment in Nepal's capital city, one marked by deep anxieties that were at once environmental, political, and cultural. At stake in Bagmati-Bishnumati restoration debates was not simply the fate of urban river reaches, but of the polity itself, and what Hansen (2001) has called the "myth of the state."

Recent scholarship on state-making has focused on the continuous production of an imaginary dimension in which the state is "a distant but persistent guarantor of a certain social order, a measure of justice, and protection from violence" (ibid.: 222). Such analyses, which recognize the layered problems of defining and studying 'the state' (Abrams 1988; Aretxaga 2003), engage it as an organizing concept through which people configure the boundaries of social cohesion, and locate the symbolic center of political will. Conceptions of nationhood, in turn, infuse the myth of the state, and, as Norvall has argued, may assume new forms during, or in response to, periods of extreme volatility or unrest (Norval 2001: 182). Experiences of deep dislocation in the wake of such volatility may also enable collective belief in the possibility of new beginnings, which, in turn, may require identities "held together as much by forgetting as by remembering" (ibid.: 187).

Over a decade when the Nepali state—as both a bureaucratic apparatus (Althusser 1971) and an assemblage of authoritative cultural ideas (Hansen and Stepputat 2001)—slowly unraveled, contests over precisely how to 'remember' a proper, ordered relationship between nature and society on the

Bagmati–Bishnumati riverscape took on a particular urgency. Embedded in the narrative of a Bagmati Civilization, then, is a complex claim to the past *and* the future, a claim through which the rivers at the heart of Nepal's capital city became terrain on which to reinvent modern urban identity and imagine wholly new politics.

This chapter traces the way Huta Ram Baidya, an adamant, outspoken advocate for the preservation of what he describes as Kathmandu Valley's cultural heritage, articulates environmental meaning through his invention and dissemination of the Bagmati Civilization concept. I aim to engage Guneratne's call, in the Introduction to this volume, to explore how human beings who live and work in Himalayan environments understand or conceptualize them, and to consider the implications of their cognized models for policy and development. Following Guneratne's point, rather than basing this analysis on more conventional components of the cultural context within which Baidya is situated, such as his ethnic, religious, or kinship position, I concentrate here on Baidya's expression of what the Bagmati River *meant* to him, and his very public efforts to diffuse that meaning.

This approach should not be interpreted as a dismissal of other elements of Baidya's cultural position; rather, it foregrounds the ways that environmental meanings can be operationalized with the expressed intention of cutting across the very identity-markers through which we often analyze cultural processes. Of considerable importance in the form and content of the Bagmati Civilization narrative, then, are the moments when environmental integrity, as both moral and political cause, is said to transcend other lines of social difference. I am interested here in the ways that the urban environment, when framed or cognized in terms of ecology or stewardship, was understood to foster social unity and cohesion where other ways of marking and practicing difference could not or did not.

The unifying, political potential of shared environmental meaning—aspirations to which are captured in landmarks of the global environmental movement such as the Bruntland Commission's *Our Common Future* (1987) and the 1992 Rio Declaration and Earth Charter—is by no means new, but it is nearly always analyzed at the global scale. In turn, global-scale assumptions about the power of ecology to transcend human social difference have been roundly and rightly critiqued for their frequent failure to "see" the very real cultural and political differences that characterize lived social life. A long legacy of scholarly attention to this global/local conundrum emphasizes, among other things, the ways that environmental improvement initiatives conceived at global scales nearly always reproduce, or newly erect, social boundaries that confound the very unity that global imaginaries seek to compel (Peters 1996; Forbes 1999; Rademacher and Patel 2002). However, analytical attention to an environmental unity narrative at a more grounded scale may shed light on the relationship between environmental meaning and related efforts to cultivate affinities across difference, most notably nationalist projects (Cederlöf and Sivaramakrishnan 2006). This is particularly true in moments of extreme political volatility—moments such as the period considered in this chapter.

The Bagmati Civilization narrative, then, presents an important opportunity to consider the place of environmental vitality narratives in broader questions regarding the state of the Nepali state and nation. As I will show, Baidya's work conceptualizes an urban environment that fosters unity across certain conventional divisions in Nepali society, even as he directly contests what he considers to be the hegemonic, and usually inept, instruments of global development, globally administered environmental improvement and management, and their attendant state institutional apparatus in Nepal. In other words, assumptions about the unifying power of the environment that may infuse international development policies and programs in Kathmandu were distinctly different in scale and composition from those that Baidya espoused through his Bagmati Civilization narrative.

Claiming an "endangered civilization for the twentieth century"

Conversations with friends and informants about the deterioration of Kathmandu's rivers often evoked strong responses inflected with nostalgia, longing, and deep anger. One morning in 2000, for example, a friend who is a tailor in Patan, told me, "the death of the Bagmati is the single greatest sorrow in my life." This very personal response to changes in the riverscape suggested that in order to understand river degradation and restoration, one had to grapple with the complex histories, imagined and real, individual and collective, that defined those terms. "I carry Bagmati in my heart and mind," another friend told me one morning, as she recalled with nostalgia her early-morning trips to Shankhamul, on the Bagmati, as a young daughter-in-law in Patan. "These days I don't visit," she said, "because I am frightened of what I'll see."

Few people seemed to have carved as intricate a public persona out of their sense of despair over river degradation as had Huta Ram Baidya, whom I first met in 1997 and subsequently revisited over the years that followed. Baidya's lifelong home in Tripureswar sits just west of the Tekucha Khola, across the street from the Bagmati's banks at Kal Mochan Temple. The gate to his compound bears a small metal sign that reads, "Huta Ram Baidya, Agricultural Engineer." He is also an inventor and a historian. Over the past decade, Baidya's views have been displayed prominently in public media, from Nepali- and English-language newspapers to Nepal Television and internet websites. In the initial stages of research on the cultural dimensions of river change in the capital, nearly everyone I consulted, from informants in the expatriate development community to Nepali academics and NGO workers, suggested that I visit Baidya; above anyone else, he was associated with public appeals for the protection of the city's waterways in what he, and others aligned with him, considered the rivers' natural and cultural entirety. During later phases of that same research, those to whom I mentioned

Baidya would often indicate their familiarity with his work, and suggest that his name had become synonymous with an idea called the "Bagmati Civilization."

In one of our earliest encounters, in October of 1997, Baidya proclaimed that he was on a "one-man crusade to save Bagmati Civilization." Over time I came to realize that his crusade entailed the promotion of a particular kind of river restoration and protection, one that involved embracing his vision of an eco-cultural whole that included not only the rivers' present status, but also the myth, pre-history, and history of an interconnected riverscape and Valley polity.

In order to explore Baidya's idea of the Bagmati Civilization, I draw on several of his key writings, as well as interviews and participant observation conducted since 1997. I begin with a discussion of his contribution to the 1995 book, *Images of a Century*, and is informed further by *The Endangered Bagmati Civilization of Kathmandu* (Baidya 1997), *Bagmati Sabhyatako Khoji* [Looking for the Bagmati Civilization] (Baidya 2001), *Bāgmati Sabhyatāko Jaggā Atikraman* [Encroaching on Bagmati Civilization Land] (Baidya 2002a), *Bāgmati Sabhyatāko Garimā* [The Importance of the Bagmati Civilization] (Baidya 2002b), and *He Bāgmati, Timi au!* [Oh Bagmati, Please Come Back!] (Baidya 2002c).[8]

Images of a Century*: the Bagmati as an indicator*

In 1995, *Images of a Century: The Changing Townscapes of the Kathmandu Valley* appeared in bookshops around Kathmandu, gaining quick popularity among development industry expatriates and middle-class urbanites for its romantic historical photos of the city. Together with a photo exhibition upon which the book was based, this volume marked a growing attention to urban cultural preservation initiatives among Nepal's elite, and within development circles. Fueled by the indisputably dizzying pace of urbanization in the Valley, preservationism was infused with a particular urgency and momentum, which in turn produced new donor initiatives based on a conviction that the city's environmental improvements should intersect with architectural preservation.[9]

Images of a Century contains a collection of photographs from 1919–92, originally compiled by the Nepal Heritage Society and a GTZ-supported NGO called Urban Development through Local Efforts. Many of the photos depict Rana[10] splendor, and a few evidence a lost urban landscape that featured an apparently cleaner, and considerably wider, Bagmati. In addition to its photographs, the book contains a collection of essays that outline urban cultural preservationist goals.

The foreword to the volume, written by editor Andreas Proksch, articulated the sense of rapid change that fueled urban development-cultural preservationist interests. In it, the Bagmati is described as a key indicator of the degradation of the cityscape, a testament to the broader cultural and environmental losses that would be incurred if countermeasures were not immediately undertaken:

> All inhabitants of Kathmandu, Lalitpur, and Bhaktapur and visitors who, at one point in their life have fallen in love with this magical Valley, have noticed the changes taking place at ever-increasing speed. Wandering through the hearts of both cities, we see nearly every day old buildings being demolished and replaced with concrete structures the size and height of which do not match those of surrounding houses. Taking a late afternoon stroll around Shankhamul Ghats, we see sewage flowing down the river where once sparkled the clear and abundant waters of the holy Bagmati. Meanwhile, the majestic skyline of the Himalaya is disappearing behind a thick layer of air pollution.
>
> (Proksch 1995: 5)

Among the essays collected in *Images of a Century* is a contribution by Huta Ram Baidya, entitled "The Holy River Bagmati." Part cultural history and part introduction to a vision of the sacred landscape, Baidya's description underlines the argument that the condition of the Bagmati should be taken as an important symptom of a broader society and polity in severe disarray.

This piece not only contributed a particular telling of history to the broader cause of cultural and architectural preservationism promoted by *Images of a Century*, it also introduced an international development and urban elite audience to Baidya's idea of a "Bagmati Civilization." In it, the Bagmati is a crucial historical site of religious training, discipline, and practice. Strict adherence to rituals of homage, purification, and the reproduction of holy sites through pilgrimage are associated with social order, continuity, and shared respect for the river. These, in turn, are related to river cleanliness and ecology, creating a nature–culture functionalism essential to the river's past vitality.

Baidya suggests that historical riverscape practices created a condition of social equity that transcended in importance an otherwise rigid social hierarchy. He notes that, "people of all castes and social status have contributed to protect, decorate, and structurally stabilize *tirthās* and riverbanks throughout the centuries." He also references the homage that modern rulers paid to the river in the form of temples and monasteries. Rather than dwelling on the terms of their rule, Baidya emphasizes the devotion that they demonstrated through the practice of temple construction. Even the infamous acts of Rana rulers like Jung Bahadur Rana[11] could be "washed away," Baidya suggests, in the act of building up the riparian templescape.

The essay also highlights Baidya's personal memories, which combine old devotional practices with a powerful sense of social obligation to the riverscape. "Mindful of public warnings," he writes of the past, "no one dared pollute the river and its surroundings." The reader is encouraged to imagine with Baidya a festive past scene of Shiva Ratri, in which he walks along a "clean, stone-paved ghat," lined by rows of willows. The implicit unity of natural order and specific cultural practices in this essay forms a key foundation of the Bagmati Civilization narrative.

In essence, "The Holy River Bagmati" traces river degradation to what Baidya calls forgetting: forgetting the river's divine origins, forgetting the social order that Baidya 'remembers' as the foundation of past socio-natural balance and regulation, and forgetting the government's "social and administrative responsibility" toward the rivers. Baidya's characterization of the present emphasizes inundation: a chaotic, unplanned influx of outsiders, the consequent overdrawing of water and accumulation of waste, and the general disaster of Valley development. The development industry, in alliance with the state, as well as the ideas of progress with which development processes are associated, are said to have exacerbated the riverscape devastation of the present.

Baidya's charge of forgotten history is not simply left to the imagination in *Images of a Century*. It is supplemented by a set of photos that include a monsoon-swollen Bagmati flanked by swimmers or worshippers (photographed in the 1920s), and the wide, shallow Bagmati of the 1940s, braided with ripples of sand.

The endangered Bagmati Civilization of Kathmandu Valley

Just before leaving Kathmandu after a period of fieldwork in 1998, I visited Huta Ram Baidya's Tripureswar home to say farewell. He greeted me with the question, "Do you know about the great Nile Civilization?" I acknowledged knowing a bit about it, and he followed with a list of others—what did I know of the Aztecs? How about the Harrappa? Telling me that these civilizations were united by their greatness and their destruction, he handed me a stack of photocopies—reproductions of his articles in Nepali and English that had recently appeared in newspapers, on the internet, and in a magazine called *The Buddha Era*. All of these pieces developed further Baidya's notion of the Bagmati Civilization; they explained what it was, where it came from, and why and how it was worth "reclaiming."

Each piece offered a familiar set of Valley origin myths, modified by placing them in a story of the rise of a great and unique civilization. Among the papers was *The Endangered Bagmati Civilization of Kathmandu Valley*, in which the following appears:

> Kathmandu Valley was once a blue, deep and beautiful lake. Legend has it that many Devas and Devis of Hindu cults and many Adi [previous manifestations of] Buddhas, were attracted to it. Bodhisatwa Jyotishree, believed to be from Manchuria, came here and by his supernatural power, cut a gorge across a cliff at Kotwal on the south of the valley to drain off this lake . . . This Bodhisatwa then established human settlement in the Valley. Jyoti-pattan was established as the capital of this

> settlement under the rule of a King. Jyotishree also gave guidelines for social development based on civics, Buddhist philosophy of life (tolerance, compassion and loving kindness to all created) and sustainable economic development. This was the first step towards the establishment of "The Bagmati Civilization."
>
> (Baidya 1997)

The essay, then, draws on a familiar origin myth, but at the same time retells it in relation to an emergent, powerful concept in international development and conservation discourse: that of sustainable development. Baidya goes on to acknowledge a host of saints, gurus, rulers, intellectuals, and devotees who moved into and out of the Valley over many centuries. He discusses the eighth-century invasion by the orthodox Hindu Sankaracharya, the raid of Samsuddhin, Sultan of Bengal (1349), and the rise to power of of Prithvi Narayan Shah, arguing that despite their initial status as outside invaders, all adapted their practices in a manner consistent with, rather than contrary to, the norms and functionality of the Bagmati Civilization. Baidya contends that in these past cases, conquest of the Valley did not constitute a violation of the Civilization's natural and cultural norms. Instead, Baidya argues, these generations of rulers showed the Civilization "respect." Of the social order that greeted all newcomers and invaders alike, Baidya writes:

> The Hindu belief that "all human beings belong to one family, created by the common Father, BRAHMA" and the Buddhist teachings of "tolerance, compassion and loving kindness" prevailed side by side among the original inhabitants. Newcomers accepted this and respected the culture of their new land. They, in their own way, contributed with their new technologies. The final outcome was peace in the Valley, and prosperity of the Bagmati Civilization . . . Indigenous culture and the social structure went through evolutionary changes. The Bagmati Civilization became richer as a major part of Nepali Civilization. This is the summary of the genesis of The Bagmati Civilization.
>
> (Baidya 1997)

The use of phrases like "original inhabitants" and "indigenous culture" imply that specifically Newar identity organized the idealized past society and ecology. Yet Baidya notably avoids giving the Bagmati Civilization an explicit ethnic identifier, foregrounding instead an organization of socio-nature that followed from a broader story of immigration, difference, and conflict subsumed by an ostensibly harmonious process of eco-cultural assimilation. This was not a conversion to indigenous identity, but rather adherence to a set of ecological norms that Baidya attaches to the Bagmati. He attributes the historical social power of the Bagmati Civilization to its capacity to accommodate newcomers and yet maintain consistency and coherence; at the same time he emphasizes so-called outsiders' willingness to adapt.

But modern history, and modern versions of urban socionature, witnessed a very different kind of immigration. In *The Endangered Bagmati Civilization of Kathmandu Valley*, Baidya suggests that the state of the riverscape marks an historical shift from immigration as adaptation to immigration as invasion. The consequence, he contends, is a disordered urban environment and massive cultural breakdown:

> The immigration still continues. Unfortunately, however, for the last three decades immigration became an influx of unmanageable dimension. Different ethnic groups from all over present day Nepal and beyond its national boundaries moved into the valley. No attempt was made by the newcomers to understand, to accept and to respect The Bagmati Civilization vision . . . The chaos thus developed is visible everywhere in the valley today. The polluted city environment and the garbage, jammed and smoky vehicular traffic, collapsing or collapsed historical buildings and temples, heritage destruction by the traditional inhabitants, by the immigrants and by the government are common happenings every day.
>
> (ibid.)

Baidya argues, then, that the contemporary condition of the Bagmati is symptomatic of an imbalance in the Valley's eco-cultural and political order. He describes the consequences:

> All rivers, including the main Bagmati, start collecting sewage, industrial pollution and poisonous chemical wastes as soon as they enter the valley floor . . . People are facing problems in their cremation formalities. Annual . . . memorial days of the deceased are no more observed on river banks. These rivers are no longer safe for ablution, and have lost their religious significance. They are now "cultural outcasts."
>
> (ibid.)

As "cultural outcasts," the rivers are said to have suffered a series of abuses. The essay describes the deleterious effects of uncontrolled sand harvesting from the riverbeds, which include sinking (Thapatali Bridge in 1991), and collapsing (Shankhamul) bridges, and caved-in stone masonry at several ghats. Baidya links sand harvesting to reinforced-concrete technology, introduced in Valley construction during the 1960s, thereby connecting modern development practices to the material experience of river "cultural abuse." Sand harvesting features in a long list of contemporary practices that have adversely impacted the riverscape's environmental integrity. Baidya names irresponsible road construction along riverbanks, park projects that encroach the river's floodplain, and prioritizing international donor interests over urban infrastructure planning as key factors in the perpetuation of abuses, presented to Valley citizens quite erroneously as development. Baidya argues, "Priority should have been given to restoring the Bagmati river system, and not on training the meandering Bagmati at random to establish a park of any name" (ibid.).

Thus for Baidya, the root of contemporary river degradation is the disruption of a relatively consistent civilizational reproduction, in which a crucial shift hinges on the forms of modernity associated with development itself. Baidya claims that it is only in the context of modernity and development that newcomers failed to respect, or take time to understand, the natural and cultural dimensions of the Valley's social order. This, combined with the persistent 'forgetfulness' of longer-established residents, brought unprecedented chaos to the riverscape and the Valley.

This essay concludes with a warning that the crisis could end in the demise of the entire historical and cultural legacy of the Bagmati Civilization. Baidya describes a finite window of opportunity for reversing the damage, calling on the people of the Valley not only to 'remember' the history he describes, but to demand better government leadership:

> If residents of the Kathmandu Valley are not sensitive to these heritage mishandlings, future efforts to restore the Bagmati Civilization will be impossible. It is time now for the government to show political commitment based on science and technology rather than get involved in political party struggles to win state power or demonstrate superiority of one political philosophy over the other under politically immature leaders. They have demonstrated that they have no political wisdom, nor do they have administrative vision and ability to run state affairs in a democratic way. If this power struggle continues, besides many other national treasures, the Bagmati Civilization will be destroyed and will be lost forever by the beginning of the 21st century just as the civilizations of the Nile of Africa . . . and Harrappa of the Indian subcontinent or Maya and Aztec of America got destroyed or got lost.
>
> (ibid.)

This is thus a political appeal with cultural-historical justifications. It suggests that an inauthentic democracy is partly to blame for environmental decline, and that more authentic democratic practices may be able to help salvage socionatural order.

Consistent with its title, *The Endangered Bagmati Civilization of Kathmandu Valley* concludes by evoking the UNESCO World Heritage Site designation attributed to several sites in the Valley.[12] Nearly seven years before UNESCO would itself make this status official,[13] Baidya deemed all of these sites endangered and culturally disrespected. Nepal's "unique credential" in the context of a "world community," signified by the idea of world heritage, is also, Baidya argues, endangered:

> Should this [destruction] happen to The Bagmati Civilization, the six UNESCO World Heritage items of the Valley will be devalued beyond

> recovery . . . It is a well-known fact that Nepal has no other unique civilization credential worth [presenting] to the world community. It is already "An Endangered Civilization of the 20th Century."
>
> (ibid.)

Thus, Baidya demands for the riverscape both a particular form of local cultural "respect" from Nepalis, and a rightful place in the prevalent global ordering of culturally valuable places that is the UNESCO World Heritage list.

As in the essay from *Images of Century, The Endangered Bagmati Civilization of Kathmandu Valley* represents degradation as a process inextricably linked to forgetting past social orders and practices that are said to have once kept the Valley's cultural and natural domains in functional balance. Both pieces locate the problem of degradation in an influx of "outsiders" said to have failed to respect a former, and now largely forgotten, order, and in the development interventions that followed. Baidya's proposed solution involves 'remembering' the past, and holding the state accountable to his notion of historical river stewardship. In this latter point, remembrance has a political articulation, mobilized in part through reflection on, and expressing dissatisfaction with, the state of the urban environment.

Additional writings, such as *Bāgmati Sabhyatāko Garimā* [The Importance of the Bagmati Civilization] and *He Bāgmati, Timi au!* [Oh Bagmati, Please Come Back!], advance arguments about the kind of socionature that the Bagmati Civilization embodies, contrasting it with the visions of nature promoted by development initiatives. In *He Bāgmati, Timi au!* he states that the development industry has tried to replace notions of nature as divine with ideas of nature that are "artificial." Park-building, such as that underway in the UN Park zone,[14] represents for Baidya an artificial built environment, while the historical Bagmati Civilization contained a more culturally consistent kind of parks. Baidya writes in *Bāgmati Sabhyatāko Garimā*:

> The river itself can be a park; there's no need to build a park . . . protect temple areas and they themselves will be parks. To develop our country and give people development, we don't have to ruin our customs and throw away our traditions. The problem with modern civilization is that we are trying to achieve development by destroying something that makes us proud.
>
> (Baidya 2002b)

Baidya further contends that the then-ubiquitous Kathmandu Municipality slogan, "Clean, Green Kathmandu" was not needed fifty years ago, before the commencement of development in the Valley. He argues in *He Bāgmati, Timi au!* that development projects encourage Kathmandu citizens to talk incessantly about how valuable the environment is, while the projects themselves bring only degradation. The more development encourages public discussions about the environment, he writes, the worse its actual condition becomes. At stake is not just a riverscape's environmental quality, but "Nepal's entire civilization."[15]

The state, development, and "cultural abuses"

Baidya's description of Bagmati Civilization history identifies two strains of forgetting: a general social forgetting of the rivers and their significance, and an administrative forgetting—a failure of the state to adhere to its river stewardship responsibilities. It is to this latter form of neglect and failure that I now turn.

Baidya was outspoken in his criticism of the relationship between the state, the development industry, and the cultural dysfunctionality that he associated with river degradation. He emphasized Kathmandu's haphazard contemporary history of urban planning in general, and planning on the rivers specifically, to directly link the dependent relationship between the Nepali state and foreign aid to river degradation.[16] Baidya argued that a true reclamation of the river system was possible only through autonomy from international development donors, and with the creation of an accountable, truly democratic, central government.

To demonstrate how development itself was a crucial culprit in urban environmental decay, Baidya regularly accompanied anyone who expressed interest in a walking tour of specific river projects and infrastructure. Included on the tour were bridges that were built over the past decade, most of which were engineered according to the degraded morphology of the riverbed at the time of construction. If river restoration was predicated on a basic need to repair and raise the deeply channelized riverbed, thereby raising river flow levels, then each new bridge constructed to degraded riverbed specifications appeared to Baidya to further limit the likelihood of restoration itself. Therefore, even as major river management plans like the Bagmati Basin Management Strategy (BBMS)[17] identified raising the riverbeds and widening the river flow as fundamental goals for restoration, actual official practices made achieving these goals both difficult and unlikely. Baidya's point was that new state-approved and development-sponsored bridge construction undermined the feasibility of officially stated restoration visions, either demonstrating an extreme lack of coordination, or bringing into question true state-development motives. He argued that such infrastructure projects spoke louder than the commitments to cultural heritage or river system improvement that were written into planning documents or espoused in official rhetoric.

Baidya also criticized the territoriality of development, noting that donor relationships seemed to allot particular development objectives—or even parts of the city—to certain donors or countries. In the late 1990s, he cynically referred to the Patan Bridge area as "Tokyo Plaza," since much of it had been built up or restored with Japanese development assistance. He was particularly frustrated with the way that the Gopal Mandir was marked after a partial restoration: a plaque was affixed in a prominent temple entranceway declaring that it had been "built through the joint efforts of the Nepal Association for the welfare of the Blind, the Tokyo Helen Keller Association with funds provided by Japan Ministry of Postal Services, Voluntary Deposit for International Aid." After a few weeks, someone scrawled a correction in black marker on the white plaque, making it read, "RE-built through the joint efforts . . ." The plaque itself remained in place, however, and was displayed far more prominently than an original inscription identifying the temple's patrons from the eighteenth century—who, of course, were also motivated by political objectives.

On several occasions, Baidya accompanied me to the site where two ghats had been destroyed to create space for the new Bagmati Bridge in 1997, and the place where a former monastery on the bridge's north end was converted into a children's park. Along the Bishnumati, he regularly pointed out the enormous public toilet buildings erected on the riverbanks as part of the Bishnumati Corridor Environmental Improvement Project. One such facility, which, like the others, drained untreated directly to the river behind it, was built just a few feet from a frequented temple site. Such development measures were, for Baidya, clear violations of Bagmati Civilization logics.

According to Baidya, Nepal's state-development alliance produced an uneven, distorted relationship between international aid, its experts and planners, and the Nepalese public; this relationship was, in turn, seen as a primary factor in river degradation. Like many others active in river politics, Baidya referred to a connection between democracy and accelerated environmental decay over the 1990s, but he reasoned that what had been referred to, and supported by, state-development interests as "democracy" was a misnomer. In reality, he argued, development and environmental policy had been profoundly undemocratic, largely dominated by international donors and their interests. Baidya pointed to blatant contradictions between what state-development policy documents said, and what was done in practice, ultimately concluding that the relationship between the state and international development institutions was itself deeply flawed. The state in this relationship was not only democratic, but perilously weak, evidenced by its failure—among other things—to enforce existing limits on the extraction of sand, contributing to the collapse of several bridges.

The Bagmati Civilization narrative carried with it, then, a suspicion of international development interventions, their institutions, and the state's relationship to them. River reclamation would require not only the revival of a social memory of Bagmati Civilization history, but would also demand state autonomy from donor influences and foreign "experts." That autonomy would have to be demanded before it would be granted, thus infusing this eco-cultural critique with a call to urgent political action.

Imagining the Bagmati Civilization in popular discourse

The idea of the Bagmati Civilization circulated widely among urban elites, activists, and state-development officials during the late 1990s. It featured in broader public discourse linking cultural loss and cultural abuse to Bagmati degradation, and became, in some ways, central to expressions of cultural heritage restoration goals in the context of urban river management.

When I walked into a photocopy shop in Kupondol in 1999, I was immediately struck by a large poster depicting a familiar photograph from *Images of a Century*: a black and white image of a monsoon-swollen Bagmati River, lined with willows. The water level in this photo was almost unimaginable in the context of current Bagmati flow patterns, except during peak flood events. Superimposed over the picture was text that had been added by the shopkeeper; it read, "Our valor! Water once flowed in the Bagmati!" When I asked the young man working in the shop where the poster came from, he replied that his father had clipped it from the newspaper, adding that he couldn't believe that the river in the photo was actually the Bagmati.

Discussions of cultural loss as a feature of Bagmati degradation appeared in various public media throughout the 1990s. Although an exhaustive account of these discussions is beyond the scope of this chapter, it is instructive to consider a small selection of examples from the popular, albeit elite English-language, press. As in Baidya's writings, references to the cultural heritage dimension of river degradation often explicitly linked riverscape decline to a broader socio-political critique. For example, a 1996 *Kathmandu Post* article entitled, "How Holy? Much Sewage Has Run Down the Bagmati," emphasized the stories of personal and cultural insult that interviewees associated with river degradation. A selection reads:

> Recalling from the past, she could not suppress the ecstasy in her eyes when she referred to the purity of the water of the Bagmati just a decade ago. "We used to bathe in the river and wash our clothes. The water was so clean that people even used it to drink. The self-centered attitude of the people, together with the deterioration in religious values, have converted the sacred river of the Hindus into an abominable gutter, making a mockery of their own cultural beliefs," Padma Kumari said.
>
> (Mainali 1996)

Later in the piece, the "self-centered attitude" identified as a catalyst for river degradation is associated with the reinstatement of democracy in 1990, which is characterized as a moment when, "like birds released from their cages, [citizens] harbored the misconception that they could do practically anything in a free world." Nepal's experience of democracy is then implicitly linked to a social disorder that gave rise to river degradation. The article goes on to outline recent assessments of water quality, and to identify some of the processes associated with river water pollution. It emphasizes the lack of coordination among various governmental regulatory bodies and the absence of a master plan for riverscape restoration. It asks, in conclusion, "is this a cultural city or an industrial one?" counter-posing the two and reinforcing a framing of degradation that relates recent development history to the complex eco-cultural plight of the city's rivers.

On April 5, 2000, a *Kathmandu Post* editorial entitled, "My House Beneath Bagmati Bridge," offered a biting critique of the Nepali experience of democracy, animated by references to uncontrolled river encroachment. In this selection, river degradation is associated with government officials who "never care whether the Bagmati came first or we did," evoking Baidya's own appeals to history and memory. A passage reads:

> This is democracy. Money is the only thing we have left to call our culture. A handful of money can hush them for sure . . . you can claim any land in this lustrous city provided you know some officials personally . . . So long as the government falls upon the hands of those who never care whether the Bagmati came first or we did, more and more houses will be built along the ever-narrowing banks of the poor Bagmati . . . And if you really have power or money in your pocket, join us. It's far more lucrative than crying out for stopping the pollution in the river.
>
> (*Kathmandu Post* 2000)

Letters and articles with titles such as, "Billions Spent but the Rivers Still Polluted" (*Kathmandu Post* 2000), "Heritage in Peril" (Baidya 2000), "My God! What a Big Drain" (Sharma 2000), and "Bagmati Waits even as Government Ignores Cleanup Call" (Subba 2000), echoed Baidya's assertion that the breakdown of river management was related to a dysfunctional state and development relationship, particularly under the Nepali experience of 1990s democracy.

Such articles, letters, and the broader public debates in which they featured, resonated with core assertions of Baidya's narrative. They linked river conditions to an absence of state autonomy and leadership, and indicted, in turn, Nepal's experience of democratization itself.

Restoration and revival

Through a narrative rendering of the lost and contemporary meanings of the capital city's rivers, Baidya itemized what would vanish if certain urban trends continued unabated. It infused those losses—both potential and realized—with explicit historical, cultural, and ecological values, and attached to their revival an imagined polity through which natural and cultural functionality could be recovered.

Specific trends were taken as consequences—even punishment—for 'forgetting' the Bagmati Civilization. These trends included a diminished generalized social respect for the rivers, a failing sense of collective social duty to steward them, and the rivers' declining utility for ritual and other meaning-making practices. This diagnosis was given an essentially political prescription—a call not only to remember, but in so doing, to hold political leaders accountable for official actions and policies associated with riverscape harm.

The catalysts for such powerful forgetting were ostensibly varied, but Baidya emphasized an era of modern development in which those who arrived as outsiders no longer respected the Civilization's sociopolitical norms. This group included development professionals and migrants, but the former were particularly implicated in a critique of an unequal relationship between the Nepali state and the development industry. Baidya saw "so-called development" as an important driving force behind Bagmati River deterioration—a primary source of cultural abuses that accelerated social forgetting and erased or effaced the material evidence of past eco-cultural order. His mission was to recover and re-teach the history of that order, and to inspire new interest in the cultural meanings and practices he associated with river revival. Citizens of a new state, born of a Bagmati Civilization remembered, might steward spaces of nature and revive the cultural integrity long missing from Nepal's modern ruling apparatus.[18]

It was through displays of "greed" that Baidya identified outsiders; his category of encroachers came to signify anyone, poor or elite, who had laid personal claim to riparian or temple land. Democracy as experienced in the 1990s was portrayed as a context in which outsiders' greed could flourish, so much so that one writer would lament, "Money is the only thing we have left to call our culture." Baidya regarded the democracy of the 1990s as inauthentic; rather than a participatory, liberating political model, he viewed it as an uncontrolled disaster for society and for the riverscape. The state had failed to deliver real democracy, Baidya argued, and the people had failed to demand it.

Yet instead of dismissing democracy as a failure, Baidya was the passionate advocate of what he considered to be a more authentic democracy, one more attuned to the social and natural particularities of the Bagmati Civilization. Despite its resonance with logics of the past,[19] Baidya's call for a more appropriate, sensitive democracy marks a significant departure from Nepal's political past, insofar as it locates the Bagmati's future in a political order that Baidya claims has never been experienced in Nepal. "Authentic," consultative democracy was a critical component of Baidya's aspirations for the riverscape, combining his appeal for the remembrance of things past with a vision of a future political order that was, at the time, without historical precedent.[20] The revival, then, was partial and selective, a simultaneous reinvention of tradition and an imagination of a never-before-experienced political order.

This chapter suggests that we take seriously the anxieties over loss that motivate the construction and diffusion of national and ecological history. These anxieties and their attendant appeals for social action illuminate how identities are fashioned and negotiated in urban environmental space, and underline the "abiding concerns of sovereignty, self-determination, and place-based collective identities [that] play out in the entanglement of nature devotion and nationalist aspirations" (Cederlöf and Sivaramakrishnan 2006: 3). Thus, the Bagmati Civilization narrative is far more than a critique of development articulated through a telling of environmental catastrophe; it is also a vision of unity afforded by a conceptualization of the urban environment that allows for some lines of social difference while rendering others irrelevant.

My intention here is by no means to suggest that the Bagmati Civilization narrative somehow erased asymmetrical power relations in the capital, or mapped authentic social equity in ways that prior models of development and environmental management did not. It is, however, to suggest that in addition to analytics of ethnicity, kinship, caste, religion, and others, we may find the environmental imaginary relevant to understanding the dynamics of identity formation, social mobilization, and the reproduction of power relations in contemporary Kathmandu.

To focus on an individual in this matter is admittedly precarious; it risks precisely the untenable radical particularism of which Guneratne writes in the Introduction to this volume. Baidya's work circulated through many layers of environmental policy, development, and activism over the period from which the data in this chapter are drawn, and an important dimension of a fuller analysis of his work and its impacts would necessarily include a detailed mapping of this circulation. For the purposes of this chapter, I have limited such a mapping to a few ethnographic moments when Baidya's influence—or, more precisely, the notion of the Bagmati Civilization—appeared to have made its way into particular kinds of consciousness about what Kathmandu's urban environment (as well as its level or lack of vitality) was, and what it was said to mean. In Guneratne's terms (this volume, pp. 3–7), it is here, in the analytical domain of cultural scales and communication, that Baidya aspired to bridge one individual's notion of the meaning of Bagmati degradation and a broader urban consciousness of the past, present, and future of a riverscape, its city, and its nation-state.

At the same time, this particular individual's evocation of the past also raises the question of Baidya's own position as an elite Newar citizen. The Newar, after all, quite famously "like to think of themselves as the indigenous people of the Kathmandu Valley" (Gellner 2003: 76). The Newar are generally regarded as the builders of the Valley's three historic royal cities: Kathmandu, Patan, and Bhaktapur, and as such all organized efforts to salvage the environmental integrity of that territory, which has stood—in different ways over time—for Newar identity even as it has also long stood as the symbolic and bureaucratic heart of the Nepali state and Shah kingship, must in some ways resonate with that ethnicity and its contemporary reproduction. While Baidya did not explicitly evoke his Newar identity in any of our conversations, or in his construction of the Bagmati Civilization narrative, there is a need for work that connects contemporary Newar identity formation to transformations in Kathmandu's built landscape and ecology, particularly given recent interest in intersections between the politics of indigeneity and environmental conservation more broadly.

This raises as well the question of audience. While, in the years during which I worked with him, he himself largely shunned the international development community as his primary audience, it was often that same community that took a strong interest in his work, ideas, and, interestingly, critiques of development itself. While Baidya himself might see his most critical sphere of communication as a Kathmandu-based Nepali audience, many outside of that sphere sought his counsel. I include myself, as both a researcher and as someone interested in the cultural dynamics of urban environmental change, in that somewhat unintended sphere.

By foregrounding Baidya's articulation of what the Bagmati River *meant*, and his efforts to diffuse that meaning, the urban environment emerges as a critical crucible in which contemporary national and city identities were forged. The Bagmati Civilization narrative was driven, in part, by its appeal to restore environmental integrity, and that integrity was defined simultaneously through moral logics and political forms. Most importantly, the narrative presented an opportunity for a "citizen" of the Civilization to declare loyalty first to a riverscape, and then to the institutions of governance that affected it. In pronouncing his own ultimate loyalty to an environmental form, Baidya suggests the possibility of unraveling historical lines of social difference and reconstituting them according to a carefully rendered environmental sovereignty.

The Bagmati Civilization narrative presents an important opportunity to consider the place of environmental vitality narratives in broader questions regarding the state of the Nepali state, the reproduction of the nation, and the particular ideologies of belonging to the capital city that circulated during a period of extreme political volatility. Through discursive construction of an urban environment that could cultivate unity across conventional divisions in Nepali society, and by pairing this with relentless political critique, Baidya offered a vision of an alternative society embedded in an alternative urban ecology. He also offered an aspiration for Kathmandu's natural and social order infused simultaneously with the reassurance of the past, and the promise of a future that would be wholly new.

Notes

1 While river degradation and general deterioration of urban environmental quality accelerated dramatically during the 1990s, the processes and discussions surrounding them have much longer genealogies. Two excellent early reviews of popular dialogue surrounding urban environmental deterioration—as well as the Bagmati and Bishnumati Rivers degradation—can be found in the journal *Himal*, May 1987 (Vol. 1, No. 1) entitled, "The Valley Chokes," and January/February 1992 (Vol. 5, No. 1) entitled, "Limits to Growth: The Weakening Spirit of Kathmandu Valley."

2 River conditions in Kathmandu have been linked to many interconnected factors. Comprehensive policy and development studies such as Stanley International (1994) identify the main causes of urban river pollution as the discharge of untreated sewage and widespread dumping of solid waste into the rivers and on their banks. Excessive sand mining in river beds and banks, which supplies mortar and cement materials to the city's construction industry, is blamed for severe morphological and flow pattern changes. In addition, most policy and development discussions of river degradation identify human encroachment on the banks, floodplains, and river beds exposed by falling water levels as an element of the degradation process. Likewise, deteriorating templescapes that line both rivers in their urban reaches are sometimes included in characterizations of river degradation.

3 This is a significance defined in popular discourse that includes, but is not limited to, devout Hindu constructions of the sacredness and power of the rivers. See Alley (2002) on Hindu perspectives on rivers.

4 Between 1996 and 2006, the Nepal government was embroiled in a bloody conflict with the Communist Party of Nepal-Maoist, which had claimed over 13,000 lives by 2006. On February 1, 2005, King Gyanendra seized direct power, which lasted until massive protests around the country forced him to reinstate Parliament in April 2006. By November 2006, the government and Maoists had signed a peace accord, bringing a provisional end to the insurgency. In April of 2008, the Maoists won the largest number of seats in elections for a new constituent assembly, and in May the new government declared Nepal a republic. Although political volatility persists at this writing, there is no doubt that Nepal has witnessed massive political transformation over the past decade. See Hutt (2004) and Thapa and Sijapati (2005).

5 Recent studies of statemaking and the environment, such as Sivaramakrishnan (1999), tend to emphasize the institutional and cultural processes that statecraft and state reproduction entail. The environment is a critical spatial and discursive terrain for the production and reproduction of state power.

6 Sivaramakrishnan (1999) points out that for Bourdieu (1990: 52), 'practice' is "the site of a dialectic between *opus operandum* and *modus operandi*, of the objectified products and incorporated products of historical practice." My consideration of cultural languages of power as part of structure and practice takes cultural meanings themselves to be "produced in social life and permeated by power relationships that are organized in time and space" (Faure and Siu 1995: 213).

7 Driver (1988) suggests this phrase as a way to analytically relate urban environments to the social production of people out of place, or outsiders. The 'moral' component marks naturalized rules or expectations that are, in actuality, contingent and constructed; it underlines the degree to which moral geographies are also ideologies of belonging. Scholarly attention to the "moral geography" of Nepal and the Kathmandu Valley through elaborations of the structure and function of sacred landscapes is exemplified by Slusser (1982), Levi (1990), Gutschow and Kolver (1975), and Gutschow *et al.* (2003).

8 This is by no means an exhaustive review of Baidya's writings, but rather a selection from the dozens of works that have outlined his ideas to a Valley audience over several years. In 2003, the Nepal Water Conservation Foundation published a collection of Baidya's writings in Nepali and English, entitled, *Bāgmati Sabhyatā Saṃrakśan Sungarśakā Bārā Barśā*, which is to date the most comprehensive collection of his written work.

9 For example, the Urban Development through Local Efforts initiative in Patan, a project of *Deutsche Gesellschaft für Technische Zusammenarbeit* (GTZ).

10 From 1846–1951, the hereditary Rana Prime Ministership ruled Nepal while Shah kings acted as figureheads. Rana rulers put large tax revenues to personal rather

than public use. *Images of a Century* evokes nostalgia powerful enough to obfuscate the repression and exploitation of the Rana Era. The book is devoted to depictions of the Ranas: court photos, hunting expeditions, Rana prime ministers, and grand interiors. Ranas are also pictured crossing the Bagmati in full regalia.

11 Jung Bahadur Kunwar, later Jung Bahadur Rana, was the first Rana ruler to rise to power following a massacre of all of his likely competitors in 1846.

12 The Kathmandu Valley World Heritage Site comprises three palaces in their urban settings (Kathmandu, Patan, Bhaktapur). Two Hindu sites (Pashupatinath and Changu Narayan), and two Buddhist sites (Swayambhunath and Boudhanath), were included in the World Heritage List in 1979.

13 UNESCO declared the Kathmandu Valley World Heritage Site officially endangered in July of 2003. See Post (2003).

14 In July of 1995, a group of acting and former state ministers, influential scholars, and development professionals proposed that a riverside park be built to commemorate the fiftieth anniversary of the United Nations (see Cosmos Engineering Services 1999: xi).

15 Notably, here the Bagmati Civilization stands for the entire nation-state of Nepal. This is consistent with historical state cultural constructions that placed the Kathmandu Valley and its cultural norms at the center of Nepaliness, or *Nepalipan*. See Rademacher (2005).

16 The development industry in Nepal was highly influential throughout the Panchayat Era, and enjoyed continued importance during the contemporary democratic period (1990–2002). See, for example, Acharya (1992), Dixit (1997), and Seddon (1987).

17 The Bagmati Basin Management Strategy (BBMS) was the main planning document of the Bagmati Basin Water Management Strategy and Investment Program. The findings of this study were released in 1994 with assistance from the Japanese Grant Fund and the World Bank, in collaboration with the Nepali Ministry of Health and Physical Planning, the International Union for the Conservation of Nature, and the consultant group Stanley International.

18 The state and Hinduism in Nepal have a long and complex history of interconnection. See, for example, Sudhindra Sharma's (2002) historicization of the relationship between statecraft and what he calls a particularly Nepali form of Hinduism.

19 An example is the establishment of the partyless Panchayat System by King Mahendra following his 1960 coup, on the grounds that it was more culturally and developmentally appropriate to Nepal than multiparty democracy. It also, notably, invested ultimate and total power in the kingship.

20 There remains at this time of writing popular enthusiasm for the new democratic order that commenced in the Spring of 2006, but it is premature to speculate as to whether *loktantra* represents the authentic democracy to which Baidya aspired.

10 Beyond cultural models of the environment

Linking subjectivities of dwelling and power

Ben Campbell

For many decades, the idea of a 'cultural model of the environment' was a valuable tool for anthropologists and other social scientists to contest the bio-physical realism of natural scientists. If we were to understand how diverse human groups interact with their specific environments, it was not enough simply to describe the objective features of those environments and human adaptation to them. We could explore what meanings people constructed of their environments, and indeed see how their categorical organisation of the natural world built into distinctive worldviews of human–environmental relationship.[1] It seemed in effect there could be no pre-cultural human response to nature.

In the 1980s, the analytical approach of culturally constructed environments acquired a certain affinity with the idea that Western constructions of the environment were about domination, profitable use, and control, whereas indigenous or traditional constructions were respectful, reciprocal and even sustainable. With the ecological sustainability agenda being increasingly addressed to ethnographic evidence, the question of how cultural models of the environment related to ecological practice assumed a new significance (Descola 1994; Baviskar 1995; Shiva 1989). At the same time, a movement in environmental anthropology that became wary of the homogenising, reifying and ethnocentric implications that the concepts of culture and nature can carry, shifted away from the frame of linguistic and symbolic systems, as ordering perceptions of the material world, to focus instead on the qualities of interaction between people and their environments (esp. Ingold 1992; Pálsson 1996). Instead of presuming that cultural categories and classifications operate deterministically on what people do and think, like some kind of software, the engagement of people with the affordances of their environment provides an alternative route into human ecological relations. This attends to the conditions in which people's skills, knowledge and intentions come about *as practices of dwelling* within environments, rather than as being applied *as cognitive designs* to environments from an exterior position of a human world separate from a bio-physical one.

There are problems and gaps in some of Ingold's theorisings, which I will explore in this chapter, especially concerning power. There is, though, a common current in critiquing nature/culture frameworks in Ingold's dwelling perspective (2000), and in political ecology (Castree and Braun 2001). The latter has de-naturalised Malthusian interpretations of population–resources dynamics to present environmental change as conditioned by structures of social inequality, rather than as gross elemental processes attributable to sheer numbers of people exploiting finite resources. With the further incorporation of discourse analysis, and its attention to the power-dimension of ways of organising and representing environmental knowledge (Brosius 1999; Leach and Fairhead 2002), a powerfully anti-essentialist tool kit for analysis of the environment and the representations social actors make of environmental processes has come about. Little of this has percolated into Himalayan anthropology. While critical studies of resource management policy exist (Ghimire 1992; Brown 1998; Kollmair *et al.* 2003), and the authority of forestry and agronomic science has been challenged by evidence of value in many indigenous practices (Messerschmidt 1990; Stevens 1993; Fisher and Gilmour 1999), there remains little theoretical work on bringing the vast literature on identity, ritual and symbolism in the Himalayas, to relate to that on the environment.

Given that the thrust of anthropological work on Nepal has been to 'de-tribalise' discussion of ethnic groups (Clarke 1980), and that over thirty years ago Sagant declared 'ethnic particularism is dead' (1976: 270), it would be a mistake to return to ethnically circumscribed worldviews for understanding environmental relations.[2] What I suggest is important is to investigate through ethnography the various claims made by social theorists, biodiversity scientists, and conservation institutions regarding human–environmental relations, and to evaluate critically the adequacy of our tools for understanding processes of environmental change, and the effects of representations concerning these processes on attempts to intervene in them. Do we find anything comparable to 'the environment' as a category used by Himalayan villagers? How are histories of human interaction with forests spoken about?[3] What claims to environmental legitimacy and control are made on the ground for local and state structures of power? What reflexivity is there among villagers of the impact of their actions on the environment, and alternatively, what are the most pressing ecologically related agendas as seen from the bottom up? I do not pretend to offer fulsome answers to all these questions in this short space, but put the questions as indicating the sorts of enquiry that are worth looking at ethnographically, and to find what people, whose interests are much spoken about, actually say in their own voices.

Perhaps more promising than the approach of culture or worldviews is that of subjectivities. In phenomenological anthropology (Viveiros de Castro 1998; Ingold 2000), in environmental discourse analysis (Leach and Fairhead 2002), in political ecology (Peluso and Watts 2001), and now in common property studies (Agrawal 2003),[4] there is a convergence on the notion that distinctive subject positions acting in environmental matters should be better understood. In contrast to the doubtful reality-status of 'cultural impact' (where are the agents, locations, collective presences, boundaries and scales of 'culture'?), a focus on subjectivities offers better possibilities for observing agency, knowledge, social interaction (including domination and resistance), experience of change, and the deliberate negotiation of relationships that have environmental effects.

'Culture', as it has been discovered by economists and environmental policy managers, risks presuming too much internal consistency, as if it were a domain or sector among others (which Williams 1961 identified as a distinctly Enlightenment notion), while the issue in the Himalaya is rather of competing dialogues, coexisting eclectic practices, and hierarchising tendencies among mutually referencing, and interpenetrating registers of belief, practice, and identity. I am not here primarily arguing with sophisticated cultural theorists, but with the unreflexive use of culture as a new policy instrument by development institutions, that are not prone to reflect on the effects of their own unspoken cultural practices on environmental outcomes (Fisher and Gilmour 1999). Yet there is a slippage from the policy instrument model of culture into some anthropological attempts to render the physical environment 'cultural', which then come to postulate culture as a meaning-generative device held by all people. This posits as a universal truth that a real physical world exists out there but we all have culturally particular apprehensions and models of it, distributed according to one's communicative socialisation.

My approach is to question the proposition that something called 'culture' operates on something else called the 'environment'. Holding these abstractions to be separate, objective and symmetrical entities is a strategy for ordering the world that has been extremely powerful in the global history of science, nature and colonialism (Drayton 2000). This gives rise today to 'systems thinking' among planners in attempts to bring domains of practice into controllable relationships,[5] and has resulted in projects to integrate cultural practice and belief (as stable and identifiable kinds of phenomena), with knowledge of biological processes.[6] The notion that culture has determining agency on people's everyday practices has led to bizarre suggestions, for example, that for purposes of conservation, the entire Himalaya could be declared a sacred grove – imagining that if declared by an appropriately sacred authority, obedience would follow. In less exaggerated form, culture has become identified as a potential tool for effectively promoting conservation initiatives. The introduction to the e-conference on 'Integrating Mountain Culture and Natural Resources Management', hosted by ICIMOD in 2001, suggested that by creating partnerships with religious and spiritual leaders, the management of natural resources and the aims of cultural integrity and survival could both be strengthened. Despite some well-nuanced analyses about problems with reifying culture (those of Kenneth Croes and Manjari Mehta in particular), such attempts to mobilise cultural inputs treat culture in the main as a substantive adjunct to conservation goals: in terms of translating one set of purposes into the language of another linguistic–cultural frame. This occurs typically within development interventions that are designed to fit outcomes and data with original project goals, without problematising the disambiguation process needed to make this fit come about (Pottier 2003). In conservation programmes, culture is a thing of knowledge rather than a problematic with which to consider epistemic transformation. When concepts such as the environment are processed from being an object of modern scientific perception, and are translated into local terminologies such as for sacred places and ethno-pharmacological knowledge, these appear in the image of corresponding to the original epistemic model. Proper cultural analysis should instead consist of being reflexive over the provisional and synthetic quality of the product of translation, and the relational-epistemic postulate that serves for thinking the same reality is being discussed.

Locating the environment

In Nepal, any attempt to deal with cultural relations with the environment is confronted by a tension between deeply particular local complexes of symbolic–material practice and the national context of development. To illustrate how my own thinking about Himalayan human–environmental relations has been influenced by fieldwork, I recall several moments when after periods spent among herders and farmers in their dispersed mountain pastures and terraces, I had my experiences of seamless, dwelt engagement with people, plants, soil and animals – in other words, deep in the stuff of Ingold's approach – abruptly confronted by encounters on the dirt road through the valley, setting villagers' environmental engagement in stark contrast to the pace of the internal combustion engine, the bazaar and bureaucracy. Were these two worlds or one? For the political ecologist within me, it is clear that the continued reliance of mountain families on subsistence production is inextricably linked to historical class and capital relations of uneven national development and peripheral neglect, yet the qualitative ethnographer, alongside, insisted on the radical disjuncture of relational universes.

Of course, similar disjunctures have been a theme of Himalayan research and analysis for the past 30 years. Holmberg (1989) wrote of the historical 'involution' of Tamang society, producing an appearance of 'tribal' features as a result of systematic exclusions from centres of cultural and political power. Burghart (1984b) wrote of how the different *des* of Nepal with their diverse environmental qualities created quite distinct worlds of value and social locatedness between *lekh, pahāṛ*, and *tarāi*. The 'complementarity' of Himalayan ecological niches was for von Fürer-Haimendorf (1975) a landscape for enterprising exchange relations and different possibilities for accumulation. Humphrey (1985) alternatively saw this as a kind of economic 'dis-integration'. Schrader (1988) analysed the quasi-autonomous economic worlds of the mountains as a system enabling the use of paternalistic community relations to limit costs for village-based elites to pursue trade with capitalist regimes of value in India. Blaikie *et al.* (1980) saw people such as the Tamang-speakers of my research, who are transhumant agro-pastoralists, as 'anarchic petty commodity producers', tangentially affected by national markets, and Sagant (1976) remarked on the nationally distributed differentiation of technologies and landscapes for irrigated rice versus highland pastoralism in Nepal. In addition to these broadly economic appraisals of differentiated Himalayan landscapes and circuits of exchange, Ramirez' (2000) study emphasises the radical political autonomy of decision-making that characterised the village universes of headmen up to the middle of the twentieth century.

Across all this literature, it seems as if local particularity, and its incompleteness, are an inescapable element of what makes up the interrelations of people and ecology. The local can never provide an adequate context of explanation, yet the structures of cultural, political, and economic integration between localities can only be understood by the role in them of local-ness. The corollary in approaching human environmental relations is that villagers' own activities of production and provisioning are not carried on in closed worlds, and have to be referenced to wider circuits, or progressive contextualisations of interaction. Yet any analysis of power relations reaching into the micro-worlds of mountain subsistence and symbolic practice is unsettled by the pervasive irrelevance of accounts that are not locally sensitive to the subjective experience of different kinds of power in actual people's lives. This problem with generality directly parallels that identified in the over-simplistic theory of Himalayan environmental degradation (Ives and Messerli 1989).

In my book (Campbell n.d.), I describe an orientation of the villagers from Rasuwa District that declares an ontology of in-betweenness. They live self-consciously in the 'middle ground' between juniper and palm tree, which is celebrated through ritual language as a golden ecology of dwelling and fertility, and yet their vertical in-betweenness is evoked in other contexts to stress the hardships of a world which is peripheral to the wealth and comforts of urban life. Subjectivities are thus framed both through extensive vertical environmental interactions, and by profiles of explicit cultural difference. Yet to speak of culturally 'mediated' environmental relations (as if culture was a cognitive enabling device on an inert material external reality), would be to displace attention from the qualities of interaction that are formed through reciprocal relations with the non-human world. It would further gloss over the significant differences within the total field of competing repertoires of relationship, knowledge and reality, which people encounter and negotiate through diverse and unstable contexts of class, kinship, ethnicity, territorial residence, and ritual in particular dynamics of history and place.

Sightings of nature–culture dualism

My research interests have been to understand the environmental relationships of Tamang-speaking communities who reside within the Langtang National Park, living on the front-line of state-sponsored, category-induced dualism between nature and society. In the prism of modernist nature protection, human presence has been perceived as inherently destructive of ecological integrity. My studies have revealed numerous ways in which this idea of nature as a non-human domain clashes with the Tamangs' modes of relationship with their environment. Approaches which do not prejudge the suitability of 'integrity' and 'interference', and use terms other than those so implicated in the situation at hand, are needed to understand the pragmatic and ontological problems the Tamang villagers face in confronting a categorical boundary-making of nature. The anthropological task of describing and analysing human–environmental relationships in these circumstances needs to interrogate rather than depend on the nature–culture opposition.

When I began fieldwork, I had expected to meet people who would produce relatively consistent narratives about the landscape, explain their 'Indigenous Environmental Knowledge', and be able to discuss matters of plants, animals, soil and weather with some reflexive facility. Instead, I found people's environmental knowledge and relationships to be discursively fragmented and not at all like a textualised body of knowledge. Their environmental practices and the contexts in which environmental factors took on relevance did not easily translate into linked strands of verbalised discourse that the ethnographer can straightforwardly access.

I had been discouraged by the fact that many villagers only knew specific names for a basic range of the most useful plants. They were reluctant to reveal local pharmacological knowledge when I asked them directly, mostly because they claimed lack of expertise of knowledge linked to ritual healing practice and occult power, though they spontaneously volunteered bits of information about the medicinal qualities [Tam. – *men daba*] of certain foods when they were cooking, like frogs and the prickly pepper spice [Tam. – *prumo*, Nep. *timur*]. They often said I would do better consulting someone else about such things, and differed significantly in their willingness to indulge my enquiries about ambient spirits and territorial gods and demons, or offer theories as to the cause of landslides. People in the village seemed too individually diverse and contradictory in their interests, activities and interpretations to justify the quest for a core cultural disposition toward the environment, beyond the basic commonalities of their dispersed, though collectively attuned, agro-pastoral production in mobile shelters [*godi*].

It was only after my initial research that I realised a fuller 'environmental' packaging was possible with the various kinds of data I had. Unlike 'kinship' which is a far more explicit and symbolically coherent language of relational identity, personhood and social practice for the Tamang, the 'environment' needed to be constructed as a synthetic analytical composite of various aspects of practice and perception in the world. There was no simple indigenously recognised entity of nature or environment, nor was there a coherent and explicit, singular cultural response to it, which certain theories of classificatory models of the environment would imply.

In the course of fieldwork days when incessant monsoon rains had cancelled arrangements for workgroups, or simply during conversations hanging out in camps of livestock shelters [*godi*], I was told oral narratives of animals and plants in mythological and everyday genres that brought alive their characteristics and habits. Later I began to translate and collate from fieldnotes various 'environmental' registers through story tellings, origination myths, and fragments of healing chants. I sought to bring these narratives into conjunction with accounts of wildlife damage to crops, and contemporary forest resource regulation issues. To give a perspective on the changing politics of environmental interactions I gathered oral histories of life before the national park when swidden agriculture, barter of forest produce, and pasture burning were common practices (now criminalised), and when the headman [*mukhiyā*] coordinated collective movements of village herds in the forest. Combined with the practical environmental intimacy of the *godi* [Nep. *goṭh*] way of life, herding animals, tending to fields, and gathering wild foods in the forest, here in these conversations was a basis for talking of an indigenous eco-relational sensibility expressed in stories of an animated landscape of interacting, diverse beings. Men, women, kings, lamas, shamans, gods, spirits, creatures, vegetation, rocks and soils featured in accounts of life process, wilful intentionality, erotic attraction, cosmic connection, species conflict, bio-type mutation, treacherous deceit, and personal fate. With the accumulating body of data came a realisation that issues of power were a constant theme in these narratives. They were very different kinds of power than that exercised by the national park authorities in claiming territorial dominion over the forest as a domain to be protected from human intervention. They constituted an indigenous ecology of power, recognition and contest in which acting subjects confronted the perspectival otherness of life kinds, inhabiting distinctive 'own worlds' of being, and entered relationships of coercion, alliance and trickery with them. Non-humans were active participants in social reality as animate beings.

Dissonances

In contrast to the view that people's cultural models and classificatory schemes provide individuals with the means to operationalise knowledge of the world in order to act upon it, I emphasise how people's relationships and interactions, their practices of dwelling, are, if not ontologically prior, at least in tension with categorical systems which are themselves mutually inconsistent and morally ambivalent. Rather than producing a coherent, culturally specific Tamang way of relating to the environment, different subject positions, agro-pastoral options, and technico-ritual dispositions – such as the knowledge specialisms of lamas, shamans [*bombo*] and territorial priests [*lhaben*] – jostle with each other, often discordantly. Differences between men and women, between clans, between village-born people and those married-in from elsewhere, all generate distinctive subjectivities of orientations and interests in environmental relationships. These differences affect how people as actors engage intentionally with their ecological circumstances, and understand their interactions as persons involved in a sociality of life forms where the social is constantly prone to cross the human/non-human divide.[7]

The mythical narrative that most clearly addressed human–environmental relationship was *Yaa Wei* (The Song of Primordial Eternity). In the sequential flow of this recitation of cosmogenesis, elements and associative patterns of ecological difference are built up to offer a grand overview of human locatedness in a tripartite cosmology of high, middle and low. Within the song, notions of vertical landscape formation and differential botanical range provide the context for the altitudinally distributed emplacement of gods, spirit creatures, and human beings. It is the tree of the middle ridge, 'sandalwood' [*surudsen*], which in the song generates a swelling of abundance. The tree's own processes of growth sustain a rooting, sprouting, branching, budding, blooming environment that affords a place to dwell for spirit residents (Kaliama Damsi Dolmo and Aba Naru Bön) and eventually their human offspring. The component parts of the tree as a structured organism are mirrored later in the part-by-part construction of a first 'golden' house in the song. That the sandalwood tree is actually unknown to its singers, except in mythical language, warns us against an over-literal interpretation of this song for understanding Tamang attitudes to actual trees, yet as with certain other myths, it is significant that a tree specified to be 'of the middle ground' becomes the means of vitalising effect for humans. In another myth of the primordial shaman Dinsur Bon, it is *shingara dongbo* (*Castanopsis tribuloides*) that plays the role of iconic middle-ground tree. This eco-ontology of mediating positionality between vertical extremes pervades Tamang language, thought and practice, and finds frequent reiteration in phrases that draw attention to 'the meeting-place of high and low' [*la deng lung ki tsam ti*].

From a phenomenological 'dwelling perspective', the *yaa wei* can be seen to sanctify a being-in-the-world of engagement and connectedness. Yet, in other cultural narratives, the in-betweenness that in *yaa wei* is a centreing of domestic existence in cosmic replenishment is turned into an in-between-ness of peripheral abandonment and poverty. On the villagers' pilgrimage up to Gosainkund for the Bhadau full moon, one of the many verses of the '*se-se bombo*' song cycle declares that the singers are not from the Tibetan town of Kyirong, nor from Kathmandu, but 'born in the middle ground, weak, unclothed and hungry'. The mid-range dwelling is here not a fortunate placement for wholesome ecological complementarity, but a marker of backwardness and misery, on an unforgiving mountainside of rocks and dark forest. It characterises the fate of middle-ground occupants as stuck in a state of powerlessness, contemplating the wealth and cultural authority of adjacent 'great traditions'. *Tserpa-i lungba* ['a miserable place'] is how villagers frequently described their homeland to me.

Similar examples of disturbances to the seamlessness of human–environmental interaction, as the dwelling perspective would like to have it, featured in conversations I had with villagers about subsistence labours, especially the care of livestock. Ingoldian seamlessness can actually be heard in many accounts of pastoral abundance, milk flowing freely, and days spent in high, flowering meadows, while at night the lights of the Kathmandu valley could be seen sparkling miles away. However, these were counterposed with bitter narratives that shifted from unreflexive, direct engagement with animals, pastures and the happy flow of time, to the abandonment many people experienced when left alone on the ridge top, running out of flour and salt to eat, having to spend hours fetching water, and forced to rely on wild spinach and plants such as rhubarb for sustenance.[8] For young wives in such situations, they imagined their husbands running around with girls down in the village. For hired herders (*goṭhālo*), it was frequently not by choice, but on account of their parents' financial debt that they spent months and years of solitary, burdensome toil, worrying that thieves, accidents on trails, and leopards would take the animals in their care, and as a result they would not receive payment in cash or cattle.

Landscapes of activity

It is only by understanding the collective commitment to mobile agro-pastoralism, that human–environmental relatedness can be understood in upper Rasuwa villages. The opposition of village and forest is not of such great significance to these Tamang-speakers as it is to fully settled cultivators. Landscape is spoken of more through the imagery of the pathway (*gyam*) than by the orientations of a *maṇḍala* model of space (Höfer 1999). Most villagers do not live primarily in the nucleated clusters of houses that characterise their village geography (for outsiders this emphasises their *jaṅgali* 'nature' or *jat*). Houses are occupied by much of the population only at the great ceremonial times of year, and for funerals. Mostly they function as storehouses. The rest of the time is spent in shifting *godi* residence between field and pasture locations, balancing the priorities of crop cultivation and surveillance, with seasonal fodder availability for livestock. Periodic movements between a household's dispersed fields break up the flow of everyday social and ecological interactions. Micro-communities of neighbourliness reconfigure themselves every month or two weeks, producing a distinctive sociality of flux and temporarily made alliances of mutual aid among new clusterings of residence, that are far more adaptable than the ideological structure of lineage solidarities (Toffin 1990).

Human conviviality with livestock induces a closeness of human–animal relations in a practical dwelling sociality of nurturance and subsistence livelihood, which sometimes even blurs ethical boundaries between humans and non-humans (Campbell 2005a). From childhood spent in the *godi*, knowledge is learnt of the tasks of looking after specific categories of animals, taking them to their best feeding grounds, and developing awareness of individual animals' desires, habits, tricks and cunning. Beyond the emotional sentiments that can arise with individual animals as characters, an understanding of the animate sociality of desires common to animals, humans and spirits of different types is developed: a common paradigm of 'giving to eat', evoked by the word *whāba*. Humans, animals, and gods-spirits [*la-lu*] all require particular attentions and have their specific dietary requirements and characteristic desires, which, if not satisfied, lead to the danger from neglected relationship of anti-social, unrequited need: a world of volatile hungry subjectivities that never know stability, but demand constant recognition and feeding for on-going sociality.

The political ecology of transhumance and forest provisioning

It emerged from conversations with older villagers that the patterns of transhumance I witnessed during fieldwork differed substantially from a generation ago, when the *mukhiyā* [village headmen] operated as semi-autocratic leaders and tax collectors (in practice till about 1960). They regulated collective movements of village cattle herds around timings of crop harvest and pasture regeneration, of sheep and goat flocks to pastures across the Tibetan border, and received fees from visiting herders of other territories. The relatively autonomous freedom of subsistence use of the forest enjoyed by villagers was contingent on relationships with the headman, involving considerable tribute in labour, and gifts of the heads of any slaughtered animals. Because of the history of an implicit moral ecology of exchange bound up in this relationship, the contemporary notion of total environmental regulation that has been introduced with the national park is problematic in its detachment from local accountability. In talking of a moral ecology, it is not to suggest there was a singular, formal cultural template of entitlements and codes of behaviour that determined how villagers and headmen pursued set roles. It seems, rather, to have been the case that the symbolism of tribute focused on the headmen, as mediators of state power and authority, was susceptible to considerable local critique for arbitrary acts of fine-imposition and excessive demands on people's labour time. A discourse from below that spoke of Tamang clans in terms of an ancient polity constituting a balance of inherited authority between kings, ministers and officials came up in conversations with older villagers as a means of symbolically contesting the singular authority conferred from above onto the old *mukhiyā*, by his annual attendance at the court of the palace at Nuwakot. By the end of the Panchayat regime, the headmen's tributary demands were frequently spoken of as a structure of rent extraction that drew livestock and labour towards the headman's own domestic enterprise. The moral ecology was thus more of a compromise between collective village practices and the performance of a model of village administration for external consumption by the state, that masked considerable ambiguity over where the extent of authority lay in everyday practice.

Subsistence rights to the resources of the forest were understood as an inherent part of belonging to the community of tax-paying households, and their performance of exacting services for the headman. Older villagers spoke of the fact that 'the *mukhiyā* said nothing' about forest product use. When, for instance, they cut timber planks and carried them three days to Kathmandu for sale, they implied he saw and took notice, but said nothing against this activity. These days with people conscious of the punitive consequences of park regulation infringements, and the power of gossip reaching officials' ears, a more limited provisioning has to be done *amrang-nale* ['without being seen'], beyond the formally licensed access to timber for projects like house-building. Local forest use regulation before the park was seen as specific to certain kinds of legitimate activity for subsistence. This did not postulate the environment as a totality in need of protection, because the very diversity of topology, plants, and animals afforded such variable uses and relationships for people, that it constituted an interactive domain of specificity rather than generality, only brought into a framing of unity as the territory of one village as opposed to another. The forest environment is still perceived as a collective good for legitimate domestic needs, but it is now inhabited by an increasing number of protected predatory wildlife, that make their unpredictable incursions on village subsistence (Campbell 2000).

From the villagers' perspectives, there was a consensual dimension for the authority of headmen that derived from functions of coordinating movements of herds, maintaining intervillage pasture boundaries, and allocating land to new settlers. This position of leadership has imploded with the national park's territorialisation of user rights (e.g. only land-owners may graze or cut fodder), and the categorical prohibitions on traditional practices such as managing ecological succession by burning, and the criminalisation of exchange of forest products without formal licensing arrangements. The crisis of legitimacy spoken of at a national scale (Gellner 2003) has its parallel at the village level in the absence of effective, consensual institutions for local subsistence accountability. Reports of the success of community forestry (Chhetri and Yonzon 1997; Fisher and Gilmour 1999; Jackson 1999) in devolving responsibility and management in much of Nepal's middle hills contrasts with the picture I have drawn of people's disempowered forest agency in the national park.

In the gulf between the state's capacity to enforce park regulations and the continuing dependence of villagers on reproducing their livelihoods through engagement with forest ecology, a more atomistic sociality of forest interaction takes place, using tactics of illicit invisibility. This more furtive forest relation hidden from public view could not be more different from reports by older villagers who recalled the collective social practice of large-scale hunting parties commanded by state officials.[9] Indeed, on one occasion when government hunters did kill a wild boar, the procession with the animal into the village centre, and its dismemberment and distribution of thighs, head and shoulders to the park and military officials, ritualised an unusual unity of purpose between villagers and state authority. Prior to park regulation, villagers additionally used to gather firewood by collective reciprocal labour groups. Such visible and publicly accountable interventions in the forest have now been outlawed, replaced in the breach by individualistic, direct engagement with the affordances of the environment. Ironically, with large-scale social forest interaction prohibited, this appears now much more like Ingold's characterisation of solitary hunter–gatherer practice, though with a glance over the shoulder for watching eyes.

To speak of 'political ecology' at the village level, it is necessary to problematise the understandings of power that relate to human–environmental interaction. For modernist administrations of biodiversity conservation, nature is explicitly a matter of territory and resources subject to a singular logic of secular control, state sovereignty and the defence of national interests, though it carries in its wake a host of inexplicit agendas of taming the cultural periphery. With the sustainability agenda, 'local environments' and 'local culture' have become privileged, but often ignoring progressive contextualisations beyond the local (Kearney 1996: 105). For the villagers, power is distributed across a multitude of life contexts, and people's accounts of forest relations tell of agency in self-provisioning, tributary recognition of entitlement granted to local leaders (Nightingale describes similar relationships in Mugu, Chapter 4 in this volume), and regulatory defiance towards state officials. People's ability to circumvent the intervention of the state in providing for domestic subsistence is one among several other concerns that require negotiating with powers that can be made to show malign and benevolent faces. In the villagers' ongoing pursuit of their extensive environmental practice they move through territories where the state is less likely to be encountered than wild animals, stray livestock, illicit grass cutters, and in recent years Maoists. The office of the Chief District Officer is one location where they occasionally go to try and alter the conditions that affect their lives, seeking to counter the authority of the national park by calling on more responsive state institutions. But many more travel the longer distance on pilgrimage to Gosainkunda or Shikar Besi to ask Mandeo (Mahadev) in person for good fortune and blessings of fertility. Their ultimate spirit of locality, Mandeo, is said to have entrapped a Newar prince as a son-in-law in an uxorilocal marriage by capture, under the waters of the high lake. This 'cultural model' of power, fantasy or delusion perhaps, sustains a challenge to the state's presumptions of territorial control, which do not attend to the making of connections with the perceived sources of environmental influence on human life. At the village scale, greater behaviour-influencing effect is noticeable in the prohibitions and taboos concerning territorial gods and spirits there (*Shyibda, lu*, etc.), and the negotiations that are made with these forces as personal interactions between animate and responsive subjectivities. Relationships with such non-secular powers contrast with the impersonal order of park bureaucracy, peopled predominantly by plains or urban Nepalis, who for the most part are unmotivated to get to know villagers' perspectives in reciprocal dialogue with them.

Conclusion

The approach I have taken towards environmental ethnography with the Tamang is that forefronting the environment has to be understood as a *synthetic manoeuvre* for perceiving diverse processes and relationships. It brings together various aspects of life-process in combinations that are not inherently held together as 'environmental' in the characteristics of their social production. Correspondingly, 'culture' is a similarly synthetic frame of analysis. The historical and cultural circumstances in which 'the environment' has been forged as a perceptual reality should not be lost sight of. The notion of the natural environment as an objectively linked whole appeared along with the revelation of its finite character, its fragile incompleteness in modern times, and the threat of its potential disappearance (Guha 1990; Grove 1993). When the environment is seen as analytically synthetic, and a historical product of human consciousness, the task of environmental anthropology becomes clearer. When brought under rigorous ethnographic scrutiny, what can be classed synthetically as environmental relationships from an outsider's standpoint (material processes and interactive relationships), decompose out of singularity and lead in many rhizomal directions beyond anything resembling a specifically 'environmental' domain. They diffuse out into social life and practice, ritual and politics. Drawing together accounts of patterns of interaction and discourses into a frame of biodiversity, requires simultaneous reflection on this process as an act of synthetic abstraction from an ethnographic totality. From an anthropological point of view, the act of abstraction that goes into producing accounts of environmental relationships needs to not forget the threadwork of social relations that give them vitality and value in the first place. Tracing associations of practices and discourses in relationships between people, plants, animals, and places has to be undertaken with an eye towards their contingency to other aspects of social reality that the Euro-American construction of the environment as an external object for the purposes of protection, does not pretend to draw into its own synthetic capacity – such as power relations.

In this task of environmental ethnography, my aim has been to identify the recognisable political-economic dimensions of transformed social relations of resource control, but in addition it has been to talk about the effects that objectifying the environment has for relations of signification regarding human personhood, collective identities, and social hierarchy in Nepal. Modern awareness of the environment as an objective and finite resource is not merely a cognitive phenomenon, that enables instrumental management, but a new element to the order of things in which historical relations of power over territory, and formations of subjectivity and difference become articulated.

In Nepal, elements of the discourse of the objectified environment and its regulation merge with caste-based ideologies of differential human closeness to natural processes through engagement with dirt, soil, sexual reproduction, and impurity. Unlike the Euro-American ideology of biologically single, universal humanity, in caste ideology humanity is radically differentiated in terms of purity–pollution that permeates all human interactions of the body with other symbolically marked bodies and with environmental elements. These terms of human differentiation operate through practices of agricultural manual labour, and carrying porters' loads, on distinctions made regarding diet (especially eating beef), and the somatic influences of local foods and water. They extend into marriage and funerary practices, and medico-ritual knowledge cumulatively distinguishing humanity into quasi-naturalised kinds.

Protecting the environment becomes appropriated within ideological dynamics of national and class projects of defining inequality in a new language. As Drayton says of British imperial science '[b]iology merely provided a new vocabulary with which to express old explanations for dominance, subordination and violence' (2000: 225). Eco-primordialism and biocentrism in conservation approaches reconfigure discourses on the legitimate residents of certain areas (Brosius 1999), with definite consequences for excluding the entitlements of marginal groups, and the ability of local communities to negotiate livelihood arrangements.

Instead of thinking about culture as a model of learned classifications that order the meanings people give to the material world of nature, Ingold argues that knowledge of the world is gained via interaction with it. If, ontologically speaking, people understand environments primarily through engaged practices of dwelling rather than through mediations of concepts, this can open up important new ways of thinking about the anthropological effects of nature conservation. When observers have noted cases of resistance to conservation, these have often been explained in terms of economic consequences for people's livelihoods (and it is through economic incentives that conservation programmes try to garner support for their projects). What the dwelling perspective makes clear is the radical ontological dissonance that can be expected by positing an objective material environment detached from human involvement, which can then be managed by conservation bureaucracies. This is most striking when Ingold discusses hunter–gatherer peoples as not being custodians of their environments in anything like the way that scientific notions of conservation based on responsible control imply: 'For hunter-gatherers this responsibility is inverted. In the last resort, it is those powers that animate the environment that are responsible for the survival or extinction of humans' (Ingold 2000: 68).

What is interesting is how the Tamang villagers resist attempts at disciplined persuasion to convince them of their powerlessness, and criminality, and continue to voice accounts of environmental relations and history which effectively place state actors as similarly motivated and compromised by their own particularistic practices as the villagers are. The processual habits and encounters of human–environmental relatedness are daily in evidence for the Tamang, who perceive their mountain environment as a field of social agency (in which villagers, the state, the supernatural, and wildlife are participant subjectivities of power), and not as a separate domain of nature in the exclusive power of others.

In her book on Eastern Tamang lyrical traditions, Steinmann writes 'Rather than being in a specifically human world, the life of people . . . inscribes them in a society in which gods, demons and animals participate, and it is to this society that universal laws apply' (2001: 280, my translation). This non-dualist view of human interaction with the world bears similarities with Viveiros de Castro's (1998) innovative perspectivist approach to humans' relations with non-humans in animist Amerindian cosmologies, and yet the Tamang live in a world of competing versions of the good life, tradition, modernity and the nation. The society of people, gods, demons and animals as Steinmann puts it, is but one possibility, and its hold on people's understanding of the world is being eroded by new careers pursued by Tamang youth eager for literacy and trekking agency jobs. The applicability of a naïve 'dwelling' perspective has to be suspect in this landscape of multiple actors, conscious of alternative regimes of value and orderings of the world. It is within this complexity and its discordant juxtaposition of interpretive schemes and valuations of practice, that the Tamang relationship with their environment has been historically located: not fully 'up' nor fully 'down', but at a median verticality where forces of attraction and exclusion operate with Tibetan and Hindu versions of the world, and where the in-between situation of both ecology and class (seasonal migrant labour) has offered a collective refuge of sorts. In this respect I diverge from Ingold's seamless engagements to draw in the active contestations Tamang villagers articulate in the positioned subjectivity of their environmental practices and representations, and to suggest their ongoing interactions with the environment have elements that are socially and politically reflexive.

In the post-Theory of Himalayan Environmental Degradation era, a limited opening to thinking about the environment as 'cultural' has been useful to try and render once absolute positions more partial and uncertain, and to listen to a more diverse set of voices (Forsyth 1998). However, within anthropological problematisations of human–environmental relatedness the analytical value of culture is an area of dispute that deserves acute theoretical and ethnographic reflection. Modernity's cosmological dividing line between the human and non-human simultaneously posits that all humans symmetrically have 'a' culture. As Viveiros de Castro (1998) puts it, this results in one nature (mute, and obeying universal laws known through science) that is perceived by a multiplicity of cultures (conveying meaning and consciousness exceptional to the human, knowable through the principle of cultural relativism). Descola (2005) has attempted to turn this insight into a typology of comparative modes of relation with the non-human, in which modernity's 'one nature, many cultures' stands structurally opposed to the cosmology of 'many natures, one culture' in the perspectivist relation of different beings to each other, that is characteristic of Amazonia, which grants all species a human-like intentionality. In his comparative typology, Descola gives totemism a separate status of its own, and a rather cumbersome residual category of 'analogism' is generated to cover the agrarian societies of Asia, Sub-Saharan Africa, the Andes and Meso-America.

While a case is made for caste ideology's analogical referencing of differential ritual status among human beings to distinct kinds of relation with the non-human, and the *maṇḍala* symbolisation of landscape indeed corresponds to an analogical frame of thinking, I suggest the ethnographic circumstances of people like the Tamang represent modes of relating to the non-human that exhibit elements of perspectivism (notable in mythological accounts of affinity among beings of different kinds), and of analogism (especially the symbolism of tribute which provides chains of connection to ritual power centres). More recently modern 'naturalism' is of course the cosmology being propagated in the form of nature conservation. Therefore, rather than seeing culture and the environment as two domains of reality that anthropologists and conservation policy thinkers are attempting to bring into relation, it is more appropriate in conditions such as pertain in the Himalayas to ask what different kinds of versions of human relationship with the non-human inform people's everyday livelihoods, their local practices of sociality, their material and discursive relations with outsiders, and their ability to negotiate with languages of power in changing significations of territoriality and citizenship.[10]

Interventions to protect the environment are now required to be reflexive over social and cultural implications and impact, and ethnographic accounts of livelihoods and forest interactions have become tools of management and policy rhetoric. In this context, it is important to think about how people understand the changing possibilities of relationship to their environments as social, and to see how relational bridges are capable of being made across the gaps of inequality between the powerful and those subject to power, when the environment becomes a principal focus of differentiation among sections of society. Under conditions of national park regulation, it is actually less possible to witness ethnographically a public and visible culture or collective community of practice in environmental relations, as state conservation imposes a mantle of protection over non-cultivated land, and defines the activities that are deemed legitimate there. National park regulations have made many people's forest interactions covert, atomistic and self-interested. In the very limited arenas for local participation with park authorities, which have emerged through buffer zone initiatives, an organisational template for a community of users accountable through management plans and committee processes does not easily translate to the internal village dynamics of lineage and party factions. Their possibilities for alliance around common interests are restricted, as the post-*mukhiyā* context of village authority has constantly struggled with a *de facto* acephalous structure, in which reversible hierarchies of kinship and affinity offer immediate constituencies of opposition to any externally 'legible' (Scott 1998) singular authority, that entities like user groups are intended to be. Alternatively, creative acts of performative solidarity with the park authorities are not entirely absent, as evidenced dramatically during my fieldwork in the common participation of villagers with the ritualised distribution of the state hunters' wild boar meat. Here a bridge of common interests and reciprocity between villagers and park guardians appeared, in which the inequality of power was validated from below.

My aim in this chapter has been to put questions of how people locate themselves as actors in places, in relationships and in problematic circumstances of livelihood. These questions were quite peripheral to the project of nature conservation until the past two decades, when policy rhetoric turned more favourably to indigenous viewpoints, and anthropologists have even dared to ask what might be the unintended consequences of regimes for environmental protection founded on scientific parameters for intervention (Fisher and Gilmour 1999: 185). However, the accounts and characterisations of local culture and knowledge favoured by conservation institutions are far from those recognised by contemporary social theory. Rather than returning to nature and culture, I suggest a focus on the ragged and unstable experiences of interaction and subjectivity in a world of differential powers will bring human–environmental relationships into a new light, and interrogate the diminished relational possibilities for human agency that attach to an environment conceived as 'out there' and under threat.

Notes

1 M. Aris (1990) eloquently describes this approach for the Himalaya, paraphrasing the words of the historian Keith Thomas (1983) concerning the classifying power that the term 'nature' assumed from the eighteenth century. But imposing this classifying characteristic on the Himalaya deserves greater attention to the effects of such a transposition across cultural and historical contexts. Thomas' argument is that 'nature' assumes a historically unique semantic and classificatory conceptual space in the European Enlightenment.

2 Concepts such as 'worldviews' can imply cultural breaks when categories such as yak-herder and urban shopkeeper are contrasted. Many people have both in the same family. 'Tibetan' patterns of polyandrous domestic economy combine such occupational differences with regularity.

3 Nightingale in Chapter 4 in this volume takes up the issue as one of seeing forests inseparably from social, cultural and political relations. Agarwal's chapter notes the instrumental possibilities of placing forests under local deities' protection, on which point Sutherland (2006) gives a remarkable account of 'rule by deity' for that same corner of the Indian Himalaya.

4 According to Agrawal (2003: 258): '[C]hanges in human subjectivities, as these

occur concomitantly with changes in institutionalized governance of the environment, are the least well understood and investigated of all environment-related changes.'

5 For example, the e-conference 'Integrating Mountain Culture and Natural Resource Management'. Online at: www.icimod.org/iym.2002/culture/mcnrm) (accessed 24 July 2001; no longer available).

6 See e.g. Shengi (1996) for a normative and functionalist view of cultural values as regulating people's environmental interactions.

7 This approach contrasts with the disaggregation of human–environmental relations into utilitarian resource interests, on one hand, and cultural models of sacred landscapes, stable values of traditional heritage, and ethno-pharmacology, on the other.

8 See March (2002: 45) for a vivid account of pastoral drudgery by a Tamang woman married to a herder.

9 Bhatt (2003) discusses the royal hunting tours up to the time of multi-party democracy, as visible performances of state power and patronage that drew park staff into close identifications with the monarchy.

10 For the Tamangs of Rasuwa, territoriality would signal the shift from their place of residence being a neglected zone of forest to becoming a national park and tourist destination. Their citizenship would have to cover their low caste status, their predominant relation to other classes of society through manual labour, and the changes in notions of collectivity brought about by Nepal's janajati movement of ethnicities, and the globalising routes of sub-continental and international labour migration.

Bibliography

Abrams, P. (1988) "Notes on the Difficulty of Studying the State," *Journal of Historical Sociology*, 1: 58–89.

Acharya, M. (1992) "Comments," in *Aid in Nepal's Development: How Necessary?* Proceedings from a Panel Debate Organized by Udaya-Himalaya Network, Kathmandu, Udaya-Himalaya Network.

Adams, V. (1998) *Doctors for Democracy*, Oxford: Oxford University Press.

Agarwal, A. (2000) "Can Hindu Beliefs and Values Help India Meet its Ecological Crisis?," in C.K. Chapple and M.E. Tucker (eds.) *Hinduism and Ecology: The Intersection of Earth, Sky and Water*, New Delhi: Oxford University Press.

Agarwal, A. and Narain, S. (1997) *Dying Wisdom: Rise, Fall and Potential of India's Traditional Water Harvesting Systems*, New Delhi: Centre for Science and Environment.

Agarwal, A., Narain, S. and Sen, S. (1999) *The Citizens' Fifth Report*, State of India's Environment, 5, New Delhi: Centre for Science and Environment.

Aggarwal, S. (2001) "Supernatural Sanctions in Commons Management: Panchayat Forest Conservation in the Central Himalayas," PhD thesis, University of Hawai'i.

Agrawal, A. (2003) "Sustainable Governance of Common-Pool Resources: Context, Methods, and Politics," *Annual Review of Anthropology*, 32: 243–62.

Aier, A. and Changkija, S. (2003) "Indigenous Knowledge and Management of Natural Resources," in T.B. Subba and G.C. Ghosh (eds.) *Anthropology of North-East India*, New Delhi: Orient Longman.

Alford, D.K. (1992) *Hydrological Aspects of the Himalayan Region*, ICIMOD Occasional Paper 18, Kathmandu: International Centre for Integrated Mountain Development.

Alley, K.D. (1998) "Idioms of Degeneracy: Assessing Gaṅga's Purity and Pollution," in L.E. Nelson (ed.) *Purifying the Earthly Body of God: Religion and Ecology in Hindu India*, New York: SUNY Press.

—— (2000) "Separate Domains: Hinduism, Politics and Environmental Pollution," in C.K. Chapple and M.E. Tucker (eds.) *Hinduism and Ecology*, New Delhi: Oxford University Press.

—— (2002) *On the Banks of the Ganga: When Wastewater Meets a Sacred River*, Ann Arbor, MI: University of Michigan Press.

Alter, J. (1992) *The Wrestler's Body: Identity and Ideology in North India*, Berkeley: University of California Press.

Althusser, L. (1971) "Ideology and Ideological State Apparatuses," in *Lenin and Philosophy and Other Essays*, New York: Monthly Review Press.

American Academy of Arts and Sciences (2001) "Religion and Ecology: Can the Climate Change?" Special issue of *Daedalus*, 130(4).

Anthwal, A., Sharma, R.C. and Sharma, A. (2006) "Sacred Groves: Traditional Way of Conserving Plant Diversity in Garhwal Himalaya, Uttaranchal," *Journal of American Science*, 2(2): 35–8.

Apffel-Marglin, F. and Parajuli, P. (2000) " 'Sacred Grove' and Ecology: Ritual and Science," in C.K. Chapple and M.E. Tucker (eds.) *Hinduism and Ecology*, New Delhi: Oxford University Press.

Aretxaga, B. (2003) "Maddening States," *Annual Review of Anthropology*, 32: 393–410.

Aris, M. (1990) "Man and Nature in the Buddhist Himalayas," in N.K. Rustomji and C. Ramble (eds.) *Himalayan Environment and Culture*, New Delhi: Indian Institute of Advanced Study.

Arnold, D. (1996) *The Problem of Nature: Environment, Culture and European Expansion*, Malden, MA: Blackwell.

Arnold, D. and Guha, R. (eds.) (1995) *Nature, Culture, Imperialism: Essays in Indian Environmental History*, Delhi: Oxford University Press.

Arora, V. (2006) "The Forest of Symbols Embodied in the Tholung Sacred Landscape of North Sikkim, India," *Conservation and Society*, 4(1): 55–83.

Avery, T.E. and Burkhart, H.E. (1994) *Forest Measurements*, 4th edn, McGraw-Hill Series in Forest Resources, New York: McGraw-Hill, Inc.

Baber, Z. (2000) "Religious Nationalism, Violence and the Hindutva Movement in India," *Dialectical Anthropology*, 25(1): 61–76.

Baidya, H.R. (1995) "Bagmati-The Holy River," in A. Proksch and H.R. Baidya (eds.) *Images of a Century: The Changing Townscapes of the Kathmandu Valley*, Kathmandu, Nepal: Deutsche Gesellschaft für Technische Zusammenarbeit: Urban Development Through Local Efforts Project.

—— (1997) "The Endangered Bagmati Civilisation of Kathmandu Valley," Online. Available at: www.users.cts.com/sd/a/abaidya/bagmati/histbagmati.com (accessed 1998).

—— (2001) "Bāgmati Sabhyatāko Khoji" (Looking for the Bagmati Civilization), *Kāntipur*, 7 Mangsir 2058.

—— (2002a) "Bāgmati Sabhyatāko Jaggā Atikraman" [Encroaching on Bagmati Civilization Land], *Kāntipur*, 28 Poush 2058.

—— (2002b) "Bāgmati Sabhyatāko Garimā" [The Importance of the Bagmati Civilization], *Kāntipur*, 2 December.

—— (2002c) "He Bāgmati, Timi au!" [Oh, Bagmati, Please Come Back!], *Kāntipur*, 22 Jeth 2058.

—— (2003) *Bāgmati Sabhyatā Saṃrakśan Sungarśakā Bārā Barśā*. Kathmandu: Nepal Water Conservation Foundation.

Baidya, J.D. (2000) "Heritage in Peril," *Kathmandu Post*, 19 February.

Bajracharya, D. (1983a) "Deforestation in the Food/Fuel Context: Historical and Political Perspectives from Nepal," *Mountain Research and Development*, 3(3): 227–40.

—— (1983b) "Fuel, Food or Forest? Dilemmas in a Nepal Village," *World Development*, 11(12): 1057–74.

Banerjee, S. (2003) "Gender and Nationalism: The Masculinization of Hinduism and

Female Political Participation in India," *Women's Studies International Forum*, 26(2): 167–79.

Bargali, S.S. (1986) "Composition of Woody Vegetation in High Elevation Blue Pine-Mixed Oak Forest, Kumaun Himalaya, India," PhD thesis, Kumaun University, Nainital, India.

Bargali, S.S., Joshi, M., Bargali, K. and Chandra, S. (1995) "Worship, Folk Art and Human Adjustment with Plant Life in Kumaun Himalayas," in B.R. Pant and M.C. Pant (eds.) *Glimpses of Central Himalaya*, Part 1, New Delhi: Radha Publications.

Barrett, C. (1996) "Fairness, Stewardship and Sustainable Development," *Ecological Economics*, 19: 11–17.

Basu, A. (1996) "Mass Movement or Elite Conspiracy? The Puzzle of Hindu Nationalism," in D. Ludden (ed.) *Making India Hindu: Religion, Community and the Politics of Democracy in India*, New Delhi: Oxford University Press.

Basu, P. and Silliman, J. (2000) "Green and Red, Not Saffron: Gender and the Politics of Resistance in the Narmada Valley," in C.K. Chapple and M.E. Tucker (eds.) *Hinduism and Ecology*, New Delhi: Oxford University Press.

Baviskar, A. (1995) *In the Belly of the River: Tribal Conflicts over Development in the Narmada Valley*, New Delhi: Oxford University Press.

—— (1999) "Vanishing Forests, Sacred Trees: A Hindu Perspective on Eco-Consciousness," *Asian Geographer*, 18(1–2): 21–31.

Beckerman, W. (1992) "Economic Growth and the Environment: Whose Growth? Whose Environment?" *World Development* 20: 481–96.

Bender, B. (ed.) (1993) *Landscape: Politics and Perspectives*, Providence, RI: Berg.

Bennett, J.W. (1976) *The Ecological Transition: Cultural Anthropology and Human Adaptation*, Oxford: Pergamon.

Berlin, I. (1981) *Against the Current: Essays in the History of Ideas*, Princeton, NJ: Princeton University Press.

Bernbaum, E. (1990) *Sacred Mountains around the World*, San Francisco: Sierra Club Books.

Bhargava, R. (2003) "Liberal Secular Democracy and Explanations of Hindu Nationalism," in A. Wyatt and J. Zavos (eds.) *Decentring the Indian Nation*, London and Portland, OR: Frank Cass.

Bhaskar, R. (1986) *Scientific Realism and Human Emancipation*, London: Macmillan and Co.

Bhatt, C. (1999) "Ethnic Absolutism and the Authoritarian Spirit," *Theory, Culture, and Society*, 16(2): 65–85.

—— (2001) *Hindu Nationalism: Origins, Ideologies and Modern Myths*, Oxford and New York: Berg.

Bhatt, N. (2003) "Kings as Wardens and Wardens as Kings: Post-Rana Ties between Nepali Royalty and National Park Staff," *Conservation and Society*, 1(2): 247–68.

Bhattacharya, S. (1970) "Kinship Terms in the Munda Languages," *Anthropos*, 65(3/4): 444–66.

Bird-David, N. (1990) "The Giving Environment: Another Perspective on the Economic System of Gatherer-Hunters," *Current Anthropology*, 31(2): 189–96.

Bishop, J. with Bishop, N. (1978) *An Ever-Changing Place: A Year among Snow Monkeys and Sherpas in the Himalayas*, New York: Simon & Schuster and Kathmandu: Centre for Action Research.

Blaikie, P. (1988) "The Explanation of Land Degradation in Nepal," in J. Ives and D.C. Pitt (eds.) *Deforestation: Social Dynamics in Watersheds and Mountain Ecosystems*, London: Routledge.

Blaikie, P. and Brookfield, H. (1987) *Land Degradation and Society*, London: Methuen.

Blaikie, P., Cameron, J., and Seddon, D. (1980) *Nepal in Crisis: Growth and Stagnation at the Periphery*, Delhi: Oxford University Press.

Blaikie, P. and Muldavin, J. (2004) "Upstream, Downstream, China, India: The Politics of Environment in the Himalayan Region," *Annals of the Association of American Geographers*, 94(3): 520–48.

Botkin, D.B. (1990) *Discordant Harmonies: A New Ecology for the Twenty-First Century*, New York: Oxford University Press.

Bourdieu, P. (1977) *Outline of a Theory of Practice*, Cambridge: Cambridge University Press.

—— (1990) *The Logic of Practice*, Stanford, CA: Stanford University Press.

Bowonder, B. (1982) "Deforestation in India," *International Journal of Environmental Studies*, 18: 223–36.

Braun, B. (2002) *The Intemperate Rainforest: Nature, Culture, and Power on Canada's West Coast*, Minneapolis: University of Minnesota Press.

Braun, B. and Castree, N. (1998) *Remaking Reality: Nature at the Millennium*, New York: Routledge.

Brosius, P. (1999) "Analyses and Interventions: Anthropological Engagements with Environmentalism," *Current Anthropology*, 40(3): 277–309.

Brower, B. (1991) *Sherpa of Khumbu: People, Livestock and Landscape*, Delhi: Oxford University Press.

Brown, K. (1998) "The Political Ecology of Biodiversity Conservation and Development in Nepal's Terai," *Ecological Economics*, 24(1): 73–87.

Bruijnzeel, L.A. and Bremmer, C.N. (1989) "Highland-Lowland Interactions in the Ganges-Brahmaputra River Basin: A Review of Published Literature," ICIMOD Occasional Paper No. 11, Kathmandu: International Centre for Integrated Mountain Development.

Bruijnzeel, L.A. and Hamilton, L.S. (2000) "Decision Time for Cloud Forests," *Humid Tropics Programme Series* no. 13, International Hydrological Program-UNESCO, Paris, IUCN-Amsterdam and WWF-Gland.

Brundtland, G.H. (1987) *Our Common Future: The World Commission on Environment and Development*, Oxford: Oxford University Press.

Brune, J.N. (1993) "The Seismic Hazard at Tehri Dam," *Tectonophysics*, 218: 281–86.

Burghart, R. (1984a) "The Tisiyahi Klinik: A Nepalese Medical Centre in an Intracultural Field of Relations," *Social Science and Medicine*, 18(7): 589–98.

—— (1984b) "The Formation of the Concept of Nation-State in Nepal," *Journal of Asian Studies*, 44(1): 101–25.

Butz, D. and Eyles, J. (1997) "Reconceptualizing Senses of Place: Social Relations, Ideology and Ecology," *Geografiska Annaler*, Series B, 79(1): 1–25.

Byers, A. (1987a) "An Assessment of Landscape Change in the Khumbu Region of Nepal Using Repeat Photography," *Mountain Research and Development*, 7(1): 77–81.

—— (1987b) "Landscape Change and Man-Accelerated Soil Loss: The Case of the Sagarmatha (Mt. Everest) National Park, Khumbu, Nepal," *Mountain Research and Development*, 7(3): 209–16.

Cain, M. (1978) "The Household Life Cycle and Economic Mobility in a Village in Bangladesh," *Population and Development Review*, 4: 421–38.

Calder, I. (2000) "Land Use Impact on Water Resources," Background Paper prepared for the Food and Agriculture Organization of the United Nations Electronic Workshop on Land-Water Linkages in Rural Watersheds, Rome, Sept.–Oct.

Callicott, J.B. (1994) *Earth's Insights: A Multicultural Survey of Ecological Ethics from the Mediterranean Basin to the Australian Outback*, Berkeley: University of California Press.

Callon, M. (1986) "Some Elements of a Sociology of Translation: Domestication of the Scallops and the Fisherman of Brieuric Bay," in J. Law (ed.) *Power, Action and Belief: A New Sociology of Knowledge*, London: Routledge and Kegan Paul.

Cameron, M.M. (1986) "Ethnomedicine and Epidemics: A Case Study of Measles in Nepal," Paper Presented at the Annual South Asia Conference, University of Wisconsin, Madison, Nov. 7–9.

—— (2009a) "Modern Desires, Knowledge Control, and Physician Resistance: Regulating Ayurvedic Medicine in Nepal," *Asian Medicine*, 4(1): 86–112.

—— (2010) Feminization and Modernization: "Women Ayurvedic Doctors and Modernizing Health Care in Nepal," *Medical Anthropology Quarterly*, 24(1).

Campbell, B. (2000) "Animals Behaving Badly: Indigenous Perceptions of Wildlife Protection in Nepal," in J. Knight (ed.) *Natural Enemies: People-Wildlife Conflict in Anthropological Perspective*, London: Routledge.

—— (2005a) "On Loving Your Water-Buffalo More than Your Own Mother," in J. Knight (ed.) *Animals in Person*, Oxford: Berg.

—— (2005b) "Nature's Discontents in Nepal," *Conservation and Society*, 3(2): 323–53.

—— (n.d.) *Living Between Juniper and Palm: Nature, Culture and Power in the Himalayas*, Manuscript in preparation.

Campbell, C.A. (1845) *The Darjeeling Guide Including a Description of the Country, and of its Climate, Soil and Productions*, Calcutta: Samuel Smith and Co.

Campbell, J.G. and Bhattarai, T.N. (1984) "People and Forests in Hill Nepal: Preliminary Presentation of Findings of Community Forestry Household and Ward Leader Survey," Community Forestry Development Project Paper 10, Kathmandu: HMG/UNDP/FAO.

Caplan, L. (1970) *Land and Social Change in East Nepal: A Study of Hindu-tribal Relations*, London: Routledge & Kegan Paul.

Carson, B. (1985) "Erosion and Sedimentation Processes in the Nepalese Himalaya," ICIMOD Occasional Paper 1, Kathmandu: International Centre for Integrated Mountain Development.

Carter, E.J. (1992) "Tree Cultivation on Private Land in the Middle Hills of Nepal: Lessons from Some Villagers of Dolakha District," *Mountain Research and Development*, 12(3): 241–55.

Castree, N. and Braun, B. (eds.) (2001) *Social Nature: Theory, Practice and Politics*, Oxford: Blackwell.

Caughley, R. (2000) *Dictionary of Chepang: A Tibeto-Burman Language of Nepal*, Canberra: Australian National University.

Cederlöf, G. and Sivaramakrishnan, K. (eds.) (2006) *Ecological Nationalisms: Nature, Livelihoods, and Identities in South Asia*, Seattle: University of Washington Press.

Chandrakanth, M.G. and Romm, J. (1991) "Sacred Forests, Secular Forest Policies and People's Actions," *Natural Resources Journal*, 31: 741–56.

Chandran, M.D.S. and Hughes, J.D. (2000) "Sacred Groves and Conservation: A Comparative Study of Traditional Reserves in the Mediterranean Area and in South India," *Environment and History*, 6: 169–86.

Chapagain, D.P. (1984) "Managing Public Lands as a Common Resource: A Village Case Study in Nepal," PhD thesis, University of Wisconsin-Madison.

Chaplin, S. (1997) "Scavengers: Still Marginalised," *Administrator*, 42: 123–51.

Chapple, C.K. and Tucker, M.E. (2000) *Hinduism and Ecology: The Intersection of Earth, Sky, and Water*, Cambridge, MA: Harvard University Press.

Chaudhury, R.R. and Rafei, U.M. (2002) *Traditional Medicine in Asia*, SEARO regional publications no. 39, New Delhi: World Health Organization (WHO), Regional Office for South-East Asia.

Chemjong, I.S. (1960) *Limbu-Nepali-English Dictionary*, Kathmandu: Nepal Akademi.

Chhetri, R., and Yonzon, P. (eds.) (1997) "Conserving Biodiversity in Nepal's Community Forests," National Biodiversity Action Plan Project: Resources Nepal, in *Proceedings of the Conference on Community Forestry in Nepal*, Kathmandu, September.

Chiriyankandath, J. (1998) "Bounded Nationalism: Kerala and the Social and Regional Limits of Hindutva," in T.B. Hansen and C. Jaffrelot (eds.) *The BJP and the Compulsions of Politics in India*, New Delhi: Oxford University Press.

Chowdhury, R.R. and Turner, B.L. (2006) "Reconciling Agency and Structure in Empirical Analysis: Smallholder Land Use in the Southern Yucatan, Mexico," *Annals of the Association of American Geographers*, 96(2): 302–22.

Christiansen, N.L. (1989) "Landscape History and Ecological Change," *Journal of Forest History*, 33(3): 116–25.

Claiborne, W. (1983) "Erosion is Laying Waste to Life in Shadow of the Himalaya," *The Washington Post*, 18 July A15.

Clarke, G. (1980) "Lama and Tamang in Yolmo," in M. Aris and Aung San Suu Kyi (eds.) *Tibetan Studies in Honour of Hugh Richardson*, Warminster: Aris and Philips.

Clements, F.E. (1936) "The Nature and Structure of the Climax," *Journal of Ecology*, 24(1): 252–84.

Cohn, B. (1990) "History and Anthropology: The State of Play," in *An Anthropologist among the Historians and Other Essays*, Delhi: Oxford University Press.

Comaroff, J. and Comaroff, J. (2006) "Introduction to *Of Revelation and Revolution*," in H.L. Moore and T. Sanders (eds.) *Anthropology in Theory: Issues in Epistemology*, Malden, MA: Blackwell Publishing.

Connell, M. (1991) *Against a Peacock Sky*, New York: Viking.

Corbridge, S.E. (1999) " 'The Militarization of all Hindudom'? The Bharatiya Janata Party, the Bomb and the Political Spaces of Hindu Nationalism," *Economy and Society*, 28(2): 222–55.

Corbridge, S.E. and Harriss, J. (2000) *Reinventing India: Liberalization, Hindu Nationalism and Popular Democracy*, Cambridge: Polity Press.

Cosmos Engineering Services (1999) *Final Report in Preparation of Conservation and Development Master Plan for Bagmati, Bishnumati-Dhobikhola Corridor*, Kathmandu: HMG Ministry of Housing and Physical Planning.

Cronin, E.W. (1979) *The Arun: A Natural History of the World's Deepest Valley*, Boston: Houghton Mifflin.

Cronon, W. (ed.) (1996) *Uncommon Ground: Rethinking the Human Place in Nature*, New York: W.W. Norton and Company.

Dahal, D.R. (1983) "Poverty or Plenty: Innovative Responses to Population Pressure in an Eastern Nepalese Hill Community," PhD thesis, University of Hawaii.

Dalton, E.T. (1872) *Descriptive Ethnology of Bengal*, Calcutta: Office of the Superintendent of Government Printing.

Daniggelis, E. (1997) *Hidden Wealth: The Survival Strategy of Foraging Farmers in the Upper Arun Valley, Eastern Nepal*, Kathmandu: Mandala Book Point.

Das, R. (2003) "Engendering Post-Colonial Nuclear Policies Through the Lens of Hindutva: Rethinking the Security Paradigm of India," *Comparative Studies of South Asia, Africa and the Middle East*, 22(1&2): 76–89.

Das, S.C. ([1902] 1970) *Journey to Lhasa and Central Tibet*, New Delhi: Manjushri.

Datta, R. (1999) "Hindu Nationalism or Pragmatic Party Politics? A Study of India's Hindu Party," *International Journal of Politics, Culture and Society*, 12(4): 573–88.

den Hertog, W.H. and Wiersum, K.F. (2000) "Timur (*Zanthoxylum armatum*) Production in Nepal: Dynamics in Nontimber Forest Resource Management," *Mountain Research and Development*, 20(2): 136–45.

Des Chene, M. (1996) "Editorial: In the Name of *Bikas*," *Studies in Nepali History and Society*, 1(2): 259–70.

Descola, P. (1994) *In the Society of Nature: A Native Ecology in Amazonia*, Cambridge: Cambridge University Press.

—— (2005) *Par-delà Nature et Culture*, Paris: Editions Gallimard.

Descola, P. and Pálsson, G. (ed.) (1996) *Nature and Society: Anthropological Perspectives*, London: Routledge.

Dhaila-Adhikari, S. and Adhikari, B.S. (2007) "Veneration of a Deity by Restoration of Sacred Grove in a Village Minar, Kumaun Region of Uttarakhand: A Case Study," *Journal of American Science*, 3(2): 45–9.

Dixit, H. (1995) *The Quest for Health*, Kathmandu, Nepal: Educational Enterprise (P.) Ltd.

Dixit, K.M. (1997) "Foreign Aid in Nepal: No Bang for the Buck," *Studies in Nepali History and Society*, 2(1): 173–86.

DN (2003) "Aid: Old Morality and New Realities," *Economic and Political Weekly*, 38(24): 2353.

Doniger, W. (1980) *Women, Androgynes, and Other Mythical Beasts*, Chicago: University of Chicago Press.

Donovan, D.G. (1981) "Fuelwood: How Much do We Need?" *Newsletter DGD 14*, Hanover, NH: Institute of Current World Affairs.

Douglas, M. (1975) "Environments at Risk," in *Implicit Meanings: Essays in Anthropology*, London: Routledge and Kegan Paul.

Dove, M.R. (1994) " 'Jungle' in Nature and Culture," in R. Guha (ed.) *Social Ecology*, Delhi: Oxford University Press.

Drayton, R. (2000) *Nature's Government: Science, Imperial Britain and the 'Improvement' of the World*, New Haven, CT: Yale University Press.

Driver, F. (1988) "Moral Geographies: Social Science and the Urban Environment in Mid-Nineteenth Century England," *Transactions of the Institute of British Geographers*, 13(3): 275–87.

Durkin, M. (1988) "Ayurvedic Treatment for Jaundice in Nepal," *Social Science and Medicine*, 27(5): 491–95.

Dwivedi, O.P. (1997) "Vedic Heritage for Environmental Stewardship," *Worldviews*, 1(1): 25–36.

—— (2000) "Dharmic Ecology," in C.K. Chapple and M.E. Tucker (eds.) *Hinduism and Ecology*, New Delhi: Oxford University Press.

Dwivedi, O.P. and Tiwari, B.N. (1999) "Environmental Protection in the Hindu Religion," in G.A. James (ed.) *Ethical Perspectives on Environmental Issues in India*, Delhi: APH Publishing Corporation.

Eck, D.L. (1982a) *Banaras, City of Light*, New York: Alfred A. Knopf.

—— (1982b) "Ganga: The Goddess in Hindu Sacred Geography," in J.S. Hawley and D.M. Wulff (eds.) *The Divine Consort: Rādhā and the Goddesses of India*, Boston, MA: Beacon Press.

Eckholm, E. (1975) "The Deterioration of Mountain Environments," *Science*, 189: 764–70.

—— (1976) *Losing Ground: Environmental Stress and World Food Prospects*, New York: W.W. Norton and Company.

Edwards, D.M. (1996) "The Trade in Non-Timber Forest Products from Nepal," *Mountain Research and Development*, 16(4): 383–94.

Eliade, M. (1963) *Myth and Reality*, New York: Harper & Row.

—— (1990) "Myths and Mythical Thought," in E. Alexander, *The Universal Myths: Heroes, Gods, Tricksters and Others*, New York: New American Library.

Ellen, R. and Fukui, K. (ed.) (1996) *Redefining Nature: Ecology, Culture, and Domestication*, Oxford: Berg.

Escobar, A. (1996) "Constructing Nature," in R. Watson and M. Peet (eds.) *Liberation Ecologies: Environment, Development, Social Movements*, London: Routledge.

Fairhead, J. and Leach, M. (1998) in L. Rival (ed.) *The Social Life of Trees*, Oxford: Berg.

Faure, D., and Siu, H.F. (eds.) (1995) *Down to Earth: The Territorial Bond in South China*, Stanford, CA: Stanford University Press.

Ferguson, J. (1994) *The Anti-Politics Machine: "Development," Depoliticization, and Bureaucratic Power in Lesotho*, Minneapolis: University of Minnesota Press.

Fickeler, P. (1962) "Fundamental Questions in the Geography of Religions," in P.L. Wagner and M.W. Mikesell (eds.) *Readings in Cultural Geography*, Chicago: University of Chicago Press.

Fisher, J.F. (1985) "The Historical Development of Himalayan Anthropology," *Mountain Research and Development*, 5(1): 99–111.

—— (1987) "Review of Fricke, *Himalayan Households: Tamang Demography and Domestic Processes*," *Himalayan Research Bulletin*, 7(2/3): 13–15.

Fisher, R. and Gilmour, D. (1999) "Anthropology and Biophysical Sciences in Natural Resource Management: Is Symbiosis Possible?" in R.B. Chhetri and O.P. Gurung (eds.) *Anthropology and Sociology of Nepal: Cultures, Societies, Ecology and Development*, Kathmandu: Sociological and Anthropological Society of Nepal.

Fisher, R.J., Singh, H.B., Pandey, D.R. and Lang, H. (1989) *The Management of Forest Resources in Rural Development: A Case Study of Sindhu Palchok and Kabhre Palanchok District of Nepal*, Mountain Populations and Institutions, Discussion Paper No. 1, Kathmandu: International Centre for Integrated Mountain Development.

Forbes, A.A. (1999) "The Importance of Being Local: Villagers, The World Bank, and NGOs in the Arun Valley, Nepal," *Identities*, 6(2–3): 319–44.

Forsyth, T. (1998) "Mountain Myths Revisited: Integrating Natural and Social Environmental Science," *Mountain Research and Development*, 18(2): 107–16.

Fortier, J. (1995) "Beyond *Jajmani*: The Complexity of Indigenous Labor Relations in Western Nepal," PhD thesis, University of Wisconsin-Madison.

—— (2009a) "The Ethnography of South Asian Foragers," *Annual Review of Anthropology*, 38: 99–114.

—— (2009b) *Kings of the Forest: The Cultural Resilience of Asian Foragers*, Honolulu: University of Hawai'i Press.

Foucault, M. (1986) "Of Other Spaces," *Diacritics*, 16(1): 22–7.

Foucault, M. and Gordon, C. (1980) *Power/Knowledge: Selected Interviews and Other Writings 1972–7*, New York: Pantheon Books.

Fox, J.M. (1984) "Firewood Consumption in a Nepali Village," *Environmental Management*, 8(3): 243–9.

—— (1993) "Forest Resources in a Nepali Village in 1980 and 1990: The Positive Influence of Population Growth," *Mountain Research and Development*, 13(1): 89–98.

Freeman, J.R. (1999) "Gods, Groves and the Culture of Nature in Kerala," *Modern Asian Studies*, 33(2): 257–302.

Fricke, T.E. (1986) *Himalayan Households: Tamang Demography and Domestic Processes*, Ann Arbor, MI: UMI Research Press.

Friends of Chipko (n.d.) *Ignoring Reason, Inviting Disaster: Threat to Ganga-Himalaya*, Dehra Dun: Friends of Chipko.

Fürer-Haimendorf von, C. (1975) *Himalayan Traders: Life in Highland Nepal*, London: John Murray.

Gadgil, M. and Chandran, M.D.S. (1992) "Sacred Groves," in G. Sen (ed.) *Indigenous Vision: Peoples of India, Attitudes to the Environment*, New Delhi: Sage Publications.

Gadgil, M. and Guha, R. (1992) *This Fissured Land: An Ecological History of India*, New Delhi: Oxford University Press.

—— (1995) *Ecology and Equity: The Use and Abuse of Nature in Contemporary India*, London and New York: Routledge.

Gaenszle, M. (1993) "Interactions of an Oral Tradition: Changes in the Muddum of the Mewahang Rai of East Nepal," in G. Toffin (ed.) *Nepal, Past and Present: Proceedings of the Franco-German Conference*, New Delhi: Sterling.

Gautam, C.M. and Watanabe, T. (2004) "Reliability of Land Use/Land Cover Assessment in Montane Nepal," *Mountain Research and Development*, 24(1): 35–43.

Geertz, C. (1963) *Agricultural Involution: The Process of Ecological Change in Indonesia*, Berkeley: University of California Press.

—— (1973) *The Interpretation of Cultures*, New York: Basic Books.

Gell, A. (1995) "The Language of the Forest: Landscape and Phonological Iconism in Umeda," in E. Hirsch and M. O'Hanlon (eds.) *The Anthropology of Landscape*, Oxford: Clarendon Press.

Gellner, D. (2003) *Resistance and the State: Nepalese Experiences*, Delhi: Social Science Press.

Gerrard, A.J. and Gardner, R. (2000) "Relationships between Rainfall and Landsliding in the Middle Hills of Nepal," *Norsk Geografisk Tidsskrift*, 54: 74–81.

—— (2002) "Relationships Between Landsliding and Land Use in the Likhu Khola Drainage Basin, Middle Hills, Nepal," *Mountain Research and Development*, 22(1): 48–55.

Gezon, L.L. (1999) "Of Shrimps and Spirit Possession: Towards a Political Ecology of Resource Management in Northern Madagascar," *American Anthropologist*, 101(1): 58–67.

Ghildial, V. and Sharma, R.C. (1985) "Some Himalayan Birds and their Conservation in Rigvedic India," *Himalayan Research and Development*, 4(2): 25–9.

Ghimire, K. (1992) *Forest or Farm: The Politics of Poverty and Land Hunger in Nepal*, Delhi: Oxford University Press.

Gilmour, D.A. (1988) "Not Seeing the Trees for the Forest: A Reappraisal of the Deforestation Crisis in Two Hill Districts of Nepal," *Mountain Research and Development*, 8(4): 343–50.

—— (1990) "Resource Availability and Indigenous Forest Management Systems in Nepal," *Society and Natural Resources*, 3: 145–58.

Gilmour, D.A., Bonell, M. and Cassells, D.S. (1987) "The Effects of Forestation on Soil Hydraulic Properties in the Middle Hills of Nepal: A Preliminary Assessment," *Mountain Research and Development*, 7(3): 239–49.

Gilmour, D.A. and Fisher, R.J. (1991) *Villagers, Forests and Foresters: The Philosophy, Processes and Practice of Community Forestry in Nepal*, Kathmandu: Sahayogi Press.

Gilmour, D.A., King, G.C., and Hobley, M. (1989) "Management of Forests for Local Use in the Hills of Nepal. 1: Changing Forest Management Paradigms," *Journal of World Forests Resource Management*, 4(2): 93–110.

Gilmour, D.A. and Nurse, M.C. (1991) "Farmer Initiatives in Increasing Tree Cover in Central Nepal," *Mountain Research and Development*, 11(4): 329–37.

Goldstein, M.C. (1983) "From a Mountain/Rural to a Plains/Urban Society: Implications of the 1981 Nepalese Census," *Mountain Research and Development*, 3(1): 61–4.

Gorer, G. ([1938] 1984) *The Lepchas of Sikkim*, Delhi: Cultural Publishing House.

Gottlieb, R. (ed.) (1996) *This Sacred Earth: Religion, Nature, Environment*, London and New York: Routledge.

Griffith, R.T.H. and Shastri, J.L. (1973) *The Hymns of the Ṛgveda*, Delhi, Motilal Banarsidass.

Grove, R. (1993) *Green Imperialism: Colonial Expansion, Tropical Island Edens, and the Origins of Environmentalism*, Cambridge: Cambridge University Press.

Grove, R., Damodaran, V. and Sangwan, S. (eds.) (1998) *Nature and the Orient: The Environmental History of South and Southeast Asia*, New Delhi: Oxford University Press.

Grove-White, R. (1993) "Environmentalism: A New Moral Discourse for Technological Society?," in K. Milton (ed.) *Environmentalism: The View from Anthropology*, London and New York: Routledge.

Guha, R. (1989) *The Unquiet Woods: Ecological Change and Peasant Resistance in the Himalaya*, Berkeley: University of California Press.

Guha, R. and Martinez-Alier, J. (1997) *Varieties of Environmentalism: Essays North and South*, London: Earthscan.

Gupta, A. and Ferguson, J. (1992) "Beyond 'Culture': Space, Identity, and the Politics of Difference," in A. Gupta and J. Ferguson (eds.) *Culture, Power, Place: Explorations in Critical Anthropology*, Durham, NC: Duke University Press.

Gupta, D. (2000) *Mistaken Modernity: India between Worlds*, New Delhi: HarperCollins Publishers India.

Gupta, P. (1999) *Swadhyaya – the Alternative Paradigm*, Mussoorie, India: Society for Integrated Development of the Himalayas.

Gurung, H. (1982) *The Himalaya: Perspective on Change*, Kathmandu: New Era.

Gururani, S. (2000) "Regimes of Control, Strategies of Access: Politics of Forest Use in the Uttrakhand Himalaya, India," in A. Agrawal and K. Sivaramakrishnan (eds.) *Agrarian Environments: Resources, Representation, and Rule in India*, Durham, NC: Duke University Press.

Guthman, J. (1997) "Representing Crisis: The Theory of Himalayan Environmental Degradation and the Project of Development in Post-Rana Nepal," *Development and Change*, 28: 45–68.

Gutschow, N. and Kolver, B. (1975) *Ordered Space: Concepts and Functions in a Town of Nepal*, Kathmandu: University Press.

Gutschow, N., Michaels, A., Ramble, C., and Steinkellner, E. (eds.) (2003) *Sacred Landscape of the Himalaya: Proceedings of an International Conference at Heidelberg*, Vienna: Austrian Academy of Sciences Press.

Gyawali, D. (1989) "Of Rabbits, Hillmen and Muddy Rivers," *Himal*, 2(2): 30–1.

Haberman, D.L. (1994) *Journey Through the Twelve Forests: An Encounter with Krishna*, New Delhi: Oxford University Press.

—— (2000) "River of Love in an Age of Pollution," in C.K. Chapple and M.E. Tucker (eds.) *Hinduism and Ecology*, New Delhi: Oxford University Press.

Hajra, P.K. (1987) "Plants in Magico-religious Beliefs and in Sanskrit Literature," in S.K. Jain (ed.) *A Manual of Ethnobotany*, Jodhpur: Scientific Publishers.

Hallpike, C.R. (1986) *The Principles of Social Evolution*, Oxford: Clarendon Press.

Hamilton, L.S. (1987) "What Are the Impacts of Himalayan Deforestation on the Ganges-Brahmaputra Lowlands and Delta?", *Mountain Research and Development*, 7(2): 256–63.

Hansen, T.B. (1996) "Recuperating Masculinity: Hindu Nationalism, Violence and the Exorcism of the Muslim 'Other'," *Critique of Anthropology*, 16(2): 137–72.

—— (1998) "The Ethics of Hindutva and the Spirit of Capitalism," in T.B. Hansen and C. Jaffrelot (eds.) *The BJP and the Compulsions of Politics in India*, New Delhi: Oxford University Press.

—— (1999) *The Saffron Wave: Democracy and Hindu Nationalism in Modern India*, New Delhi: Oxford University Press.

—— (2001) "Governance and State Mythologies in Mumbai," in T.B. Hansen and F. Stepputat (eds.) *States of Imagination: Ethnographic Explorations of the Postcolonial State*, Durham, NC: Duke University Press, pp. 221–56.

Hansen, T.B. and Stepputat, F. (2001) *States of Imagination: Ethnographic Explorations of the Postcolonial State*, Durham, NC: Duke University Press.

Haraway, D.J. (1997) *Modest_Witness@Second_Millennium. FemaleMan©_Meets _OncoMouse*176, New York: Routledge.

Hardman, C. (1981) "The Psychology of Conformity and Self-Expression among the Lohorung Rai of East Nepal," in P. Heelas and A. Lock (eds.) *Indigenous Psychologies: The Anthropology of the Self*, London: Academic Press.

Harré, R. and Bhaskar, R. (2005) "How to Change Reality: Story v. Structure. A Debate between [Rom Harré and Roy Bhaskar]," in J. Lopez and G. Potter (eds.) *After Postmodernism: An Introduction to Critical Realism*, New York: Continuum International Publishing Group.

Harris, M. (1979) *Cultural Materialism: The Struggle for a Science of Culture*, New York: Vintage Books.

Harriss-White, B. (2003) *India Working: Essays on Society and Economy*, Cambridge: Cambridge University Press.

Hawkins, T. (1981) *Community Forestry Development in Dhaulagiri Forest Division,*

1980–1981, Miscellaneous Document No. 6, NEP/80/030, Kathmandu: HMG/UNDP/FAO Community Forestry Development Project.

Henry, J.D. and Swan, J.M. (1974) "Reconstructing Forest History from Live and Dead Plant Material – An Approach to the Study of Forest Succession in Southwest New Hampshire," *Ecology*, 55(4): 772–83.

Himalayan Ayurveda Research Institute (1996) *Project Report: Medicinal Plants and Traditional Medical Practice in Gorkha District*, Kathmandu, Nepal: Gorkha Development Project/GTZ.

Hirsch, E. and O'Hanlon, M. (eds.) (1995) *The Anthropology of Landscape: Perspective on Place and Space*, Oxford: Clarendon Press.

His Majesty's Government of Nepal (1999) *Forest and Shrub Cover of Nepal 1994*, Forest Survey Division, Department of Forest Research and Survey, Ministry of Forest and Soil Conservation, Forest Resource Information System Project/Government of Finland, Publication 72, Kathmandu.

Hobart, M. (ed.) (1993) *An Anthropological Critique of Development: The Growth of Ignorance*, London and New York: Routledge.

Hoben, A. (1996) "The Cultural Construction of Environmental Policy: Paradigms and Politics in Ethopia," in M. Leach and R. Mearns (eds.) *The Lie of the Land: Challenging Received Wisdom on the African Environment*, Oxford: International African Institute.

Hobley, M. (1996) "The Four Ages of Indian Forestry: Colonialism, Commercialism, Conservation, and Collaboration," in M. Hobley (ed.) *Participatory Forestry: The Process of Change in India and Nepal*, London: Rural Development Forestry Network, Overseas Development Institute.

Hobley, M. and Malla, Y.B. (1996) "From Forests to Forestry—The Three Ages of Forestry in Nepal: Privatization, Nationalization, and Populism," in M. Hobley (ed.) *Participatory Forestry: The Process of Change in India and Nepal*, London: Rural Development Forestry Network, Overseas Development Institute.

Hobsbawm, E. and Ranger, T. (1983) *The Invention of Tradition*, Cambridge: Cambridge University Press.

Höfer, A. (1999) "*Nomen est numen*: Notes on the Verbal Journey in Some Western Tamang Oral Ritual Texts," in B. Bickel and M. Gaenszle (eds.) *Himalayan Space: Cultural Horizons and Practices*, Zürich: Völkerkundemuseum.

Holmberg, D. (1989) *Order in Paradox: Myth, Ritual, and Exchange among Nepal's Tamang*, Ithaca, NY: Cornell University Press.

Howell, S. (1996) "Nature in Culture, or Culture in Nature? Chewong Ideas of Humans and Other Species," in P. Descola and G. Pálsson (eds.) *Nature and Society: Anthropological Perspectives*, London: Routledge.

Huber, T. (1999) *The Cult of Pure Crystal Mountain: Popular Pilgrimage and Visionary Landscape in Southeast Tibet*, Oxford: Oxford University Press.

Humphrey, C. (1985) "Barter and Economic Disintegration," *Man*, 20(1): 48–72.

—— (1995) "Chiefly and Shamanist Landscapes in Mongolia," in E. Hirsch and M. O'Hanlon (eds.) *The Anthropology of Landscape*, Oxford: Clarendon Press.

Hutt, M. (2001) *Devkota's Muna Madan: Translation and Analysis*, Lalitpur, Nepal: Sajha Prakashan.

—— (2004) *Himalayan People's War: Nepal's Maoist Rebellion*, Bloomington, IN: Indiana University Press.

IDS (1983) *Foreign Aid and Development in Nepal*, Kathmandu: Integrated Development Systems.

Indian Express (1999) "Tehri Will Have To Go: VHP," 13 April.

Ingles, A.W. (1997) "The Influence of Religious Beliefs and Rituals on Forest Conservation in Nepal," in K. Seeland (ed.) *Nature is Culture: Indigenous Knowledge and Socio-Cultural Aspects of Trees and Forests in Non-European Cultures*, London: Intermediate Technology Publications.

Ingold, T. (1986) *The Appropriation of Nature: Essays on Human Ecology and Social Relations*, Manchester: Manchester University Press.

—— (1992) "Culture and the Perception of the Environment," in E. Croll and D. Parkin (eds.) *Bush Base: Forest Farm: Culture, Environment and Development*, London and New York: Routledge.

—— (2000) *Perception of the Environment: Essays in Livelihood, Dwelling and Skill*, London: Routledge.

International Rivers Network (2002) "Fact Sheet: Tehri Dam." Online. Available at: www.irn.org/programs/india/021022.tehrifactsheet.pdf.

Isaac, E. (1967) "God's Acre—Property in Land: A Sacred Origin?" *Landscape*, 14: 28–32.

Ives, J.D. (1980) "Crisis in the Himalaya," *Development Forum*, 9(7): 9–10.

—— (1986) "The Status of Nepal: Development in the Face of Uncertainty," in S.C. Joshi (ed.) *Nepal Himalaya: Geo-Ecological Perspectives*, Nainital: Himalayan Research Group.

—— (1987) "The Theory of Himalayan Environmental Degradation: Its Validity and Application Challenged by Recent Research," *Mountain Research and Development*, 7(3): 189–99.

—— (1999) "The Use of Hillside Environments for Land Husbandry: Personal reflections," *Mountain Research and Development*, 19(3): 173–7.

—— (2004) *Himalayan Perceptions*, New York: Routledge.

Ives, J.D. and Messerli, B. (1989) *The Himalayan Dilemma: Reconciling Development and Conservation*, Tokyo: The United Nations University and London and New York: Routledge.

Ives, J.D., Messerli, B. and Thompson, M. (1987) "Research Strategy for the Himalayan Region: Conference Conclusions and Overview," *Mountain Research and Development*, 7(3): 332–44.

Jackson, C. and Chattopadhyay, M. (2001) "Identities and Livelihoods: Gender, Ethnicity and Nature in a South Bihar Village," in A. Agrawal and K. Sivaramakrishnan (eds.) *Social Nature: Resources, Representations and Rule in India*, New Delhi: Oxford University Press.

Jackson, W. (1999) "The Dynamics of *Lekh* Forest Use in the Middle Hills of Nepal," DPhil thesis, University of Western Sydney.

Jackson, W.J., Tamrakar, R.M., Hunt, S. and Shepherd, K.R. (1998) "Land-use Changes in Two Middle Hills Districts of Nepal," *Mountain Research and Development*, 18(3): 193–212.

Jaffrelot, C. (1996) *The Hindu Nationalist Movement and Indian Politics, 1925 to the 1990s*, London: Hurst & Co.

—— (2003) *India's Silent Revolution: The Rise of the Lower Castes*, London: Hurst & Co.

James, G.A. (ed.) (1999) *Ethical Perspectives on Environmental Issues in India*, Delhi: APH Publishing Corporation.

Jeffery, R. (1998) *The Social Construction of Indian Forests*, New Delhi: Manohar.

Jenkins, R. (1999) *Democratic Politics and Economic Reform in India*, Cambridge: Cambridge University Press.

Jha, D.N. (1999) "Against Communalising History," General President's Address, Punjab History Conference, Patiala, March.

—— (2002) *The Myth of the Holy Cow*, London: Verso.

Jodha, N.S. (1995) "The Nepal Middle Mountains," in J.X. Kasperson, R.E. Kasperson and B.L. Turner (eds.) *Regions and Risk: Comparisons of Threatened Environments*, Tokyo: United Nations University.

Johnson, A. (1974) "Ethnoecology and Planting Practices in a Swidden Agricultural System," *American Ethnologist*, 1(1): 87–101.

Johnson, K., Olson, E.A. and Manandhar, S. (1982) "Environmental Knowledge and Response to Natural Hazards in Mountainous Nepal," *Mountain Research and Development*, 2(2): 175–88.

Joshi, S. (2001) *Fractured Modernity: Making of a Middle Class in Colonial North India*, New Delhi: Oxford University Press.

Justice, J. (1986) *Policies, Plans and People: Culture and Health Development in Nepal*, Berkeley: University of California Press.

Kaith, D.C. (1960) "Forest Practices in Control of Avalanches, Floods, and Soil Erosion in the Himalayan Front," Vol. III: *Fifth World Forestry Congress Proceedings*, Seattle.

Kanel, K.R. (2000) "Sustainable Management of Medicinal and Aromatic Plants in Nepal: A Strategy," Study Commissioned by IDRC/SARO, Medicinal and Aromatic Plants Program in Asia (MAPPA), New Delhi.

Kathmandu Post (2000a) "My House Beneath Bagmati Bridge" (editorial), April 5.

—— (2000b) "Billions Spent but the Rivers Still Polluted" (editorial), February 7.

Katju, M. (2003) *Vishva Hindu Parishad and Indian Politics*, Delhi: Sangam Books.

Kearney, M. (1996) *Reconceptualizing the Peasantry*, Boulder, CO: Westview Press.

Keesing, R. (1987) "Anthropology as Interpretive Quest," *Current Anthropology*, 28(2): 161–76.

Kelkar, G. and Nathan, D. (1991) *Gender and Tribe: Women, Land and Forests*, London: Sage Publications.

Kendall, L. (1998) "Who Speaks for Korean Shamans when Shamans Speak of the Nation?" in D. Gladney (ed.) *Making Majorities: Constituting the Nation in Japan, Korea, China, Malaysia, Fiji, Turkey, and the United States*, Stanford, CA: Stanford University Press.

Khanna, M. (2000) "The Ritual Capsule of Durga Puja: An Ecological Perspective," in C.K. Chapple and M.E. Tucker (eds.) *Hinduism and Ecology*, New Delhi: Oxford University Press.

Kharel, S. (1993) "Women's Participation in Community Forestry: The Nepal-Australia Community Forestry Project's Experience," *Banko Janakari*, 4: 73–75.

Kishwar, M. (1992) "Demanding an Honest, Open Review: Movement against the Tehri Dam," *Manushi*, 70: 2–21.

—— (1995) "A Himalayan Catastrophe: The Controversial Tehri Dam in the Himalayas," *Manushi*, 91: 5–16.

Kohli, A. (1990) *Democracy and Discontent: India's Growing Crisis of Governability*, Cambridge: Cambridge University Press.

Kohn, T. (1995) " 'She Came Out of the Field and Into My Home': Reflections,

Dreams and a Search for Consciousness in Anthropological Method," in A.P. Cohen and N. Rapport (eds.) *Questions of Consciousness*, London: Routledge.

Kollmair, M. and Müller-Böker, U. (2002) "Forest and Trees: Changing Perspectives on a Natural Resource in the Nepal Himalaya," *Global Environmental Research*, 6(1): 73–84.

Kollmair, M., Müller-Böker, U. and Soliva, R. (2003) "The Social Context of Nature Conservation in Nepal," *European Bulletin of Himalayan Research*, 24: 25–62.

Kondos, A. (1987) "The Question of 'Corruption' in Nepal," *Mankind*, 17(1): 15–29.

Korom, F.J. (1998) "On the Ethics and Aesthetics of Recycling in India," in L.E. Nelson (ed.) *Purifying the Earthly Body of God*, New York: SUNY Press.

Kuhn, T. (1962) *The Structure of Scientific Revolutions*, 2nd edn., Chicago: University of Chicago Press.

Lachapelle, P., Smith, P.D. and McCool, S.F. (2004) "Access to Power or Genuine Empowerment? An Analysis of Three Community Forest Groups in Nepal," *Human Ecology Review*, 11(1): 1–12.

Lakha, S. (2000) "The State, Globalisation and Indian Middle Class Identity," in B.H. Chua (ed.) *Consumption in Asia: Lifestyles and Identities*, London: Routledge.

Langford, J.M. (2002) *Fluent Bodies: Ayurvedic Remedies for Postcolonial Imbalance*, Durham, NC: Duke University Press.

Latour, B. (1997) "Constructing Actor-Network Theory," in M. Michael (ed.) *Constructing Identities*, London: Sage.

—— (2004) "Why Has Critique Run Out of Steam? From Matters of Fact to Matters of Concern," *Critical Inquiry*, 30: 225–48.

Lawoti, S. (2005) "Nepal's Last Hunter-Gatherers: Time is Running out for the Raute and Their Nomadic Existence," *Nepali Times* (*Nation* Section) June 17–24. Online. Available at: www.thdl.org/texts/reprints/nepali_times/Nepali_Times_252.pdf.

Leach, M. and Fairhead, J. (2002) "Anthropology, Culture and Environment," in J. MacClancy (ed.) *Exotic No More: Anthropology on the Front Lines*, Chicago: University of Chicago Press.

Leach, M. and Mearns, R. (1996a) *The Lie of the Land: Challenging Received Wisdom on the African Environment*, Oxford: International African Institute.

—— (1996b) "Environmental Change and Policy: Challenging Received Wisdom in Africa," in M. Leach and R. Mearns (eds.) *The Lie of the Land*, Oxford: International African Institute.

Lee, R.B. (1968) "What Hunters Do for a Living, or, How to Make Out on Scarce Resources," in R.B. Lee and I. DeVore (eds.) *Man the Hunter*, Chicago: Aldine Publishing Co.

Levenson, B. (1979) *Fuelwood Utilization: A Study of the Demand and Availability of Fuelwood Resources at Six Selected Villages*, Kathmandu: Integrated Water Management Project (IWM/PTR/9) Ministry of Forest, Department of Soil and Water Conservation, HMG Nepal.

Levi, R. (1990) *Mesocosm: Hinduism and the Organization of a Traditional Newar City in Nepal*, Berkeley: University of California Press.

Li, T. (1999) "Compromising Power: Development, Culture, and Rule in Indonesia," *Cultural Anthropology*, 14(3): 295–322.

Limbu, D.B. (1989) *Kirāt Paramparā, Limbu Riti-Thiti rā Saṃskrti*, Mirik: Kirat Chumlung Sabha.

Lopez, J. and Potter, G. (2005) "After Postmodernism: The New Millennium," in

J. Lopez and G. Potter (eds.) *After Postmodernism: An Introduction to Critical Realism*, New York: Continuum International Publishing Group.

Low, A. and Tremayne, S. (2001) "Introduction," in A. Low and S. Tremayne (eds.) *Women as Sacred Custodians of the Earth? Women, Spirituality and the Environment*, New York: Berghahn Books.

Ludden, D. (ed.) (1996) *Making India Hindu: Religion, Community, and the Politics of Democracy in India*, Delhi: Oxford University Press.

Lutgendorf, P. (2000) "City, Forest, and Cosmos: Ecological Perspectives from the Sanskrit Epics," in C.K. Chapple and M.E. Tucker (eds.) *Hinduism and Ecology*, New Delhi: Oxford University Press.

Maag, B. (1997) "Forests and Trees in the World of Two Tamang Villages in Central Nepal: Observations with Special Reference to the Role of Tamang Women in Forest Management," in K. Seeland (ed.) *Nature Is Culture: Indigenous Knowledge and Socio-Cultural Aspects of Trees and Forests in Non-European Cultures*, London: Intermediate Technology Publications, Ltd.

Macfarlane, A. (1968) "Population Crisis: Anthropology's Failure," *New Society*, 315, 10 October.

—— (1976) *Resources and Population: A Study of the Gurungs of Nepal*, Cambridge: Cambridge University Press.

—— (2001) "Sliding Downhill: Some Reflections on Thirty Years of Change in a Himalayan Village," *European Bulletin of Himalayan Research*, 20–1: 105–24.

Madan, T.N. (1987 "Secularism in its Place," *Journal of Asian Studies*, 46(4): 747–59.

Mahānāma, Geiger, W. and Bode, M. H. (trans.) ([1912] 1986) *The Mahāvaṃsa, or, The Great Chronicle of Ceylon*, New Delhi: Asian Educational Services.

Mahapatra, R. (2001) "Betrayed: Nepal's Forest Bureaucracy Prepares for the Funeral of the Much Hailed Community Forest Management Programme," *Down To Earth*, 9(22): 20.

Mahat, T.B.S., Griffin, D.M. and Shepherd, K.R. (1986a) "Human Impact on Some Forests of the Middle Hills of Nepal Part 1. Forestry in the Context of the Traditional Resources of the State," *Mountain Research and Development*, 6(3): 223–32.

—— (1986b) "Human Impact on Some Forests of the Middle Hills of Nepal Part 2. Some Major Impacts Before 1950 on the Forests of Sindhu Palchok and Kabhre Palanchok," *Mountain Research and Development*, 6(4): 325–34.

—— (1987a) "Human Impact on Some Forests of the Middle Hills of Nepal Part 3. Forest in the Subsistence Economy of Sindhu Palchok and Kabhre Palanchok," *Mountain Research and Development*, 7(1): 53–70.

—— (1987b) "Human Impact on Some Forests of the Middle Hills of Nepal Part 4. A Detailed Study in Southeast Sindhu Palchok and Northeast Kabhre Palanchok," *Mountain Research and Development*, 7(2): 111–33.

Mainali, A. (1996) "How Holy? Much Sewage Has Run Down the Bagmati," *Kathmandu Post*, n.d.

Malinowski, B. (1926) *Myth in Primitive Psychology*, New York: W.W. Norton and Co.

Malla, Y.B. (2001) "Changing Policies and the Persistence of Patron–Client Relations in Nepal: Stakeholders' Responses to Changes in Forest Policies," *Environmental History*, 6(2): 287–307.

Malla, Y.B., Neupane, H.R. and Branney, P.J. (2003) "Why Aren't Poor People Benefiting More from Community Forestry?" *Journal of Forest and Livelihood*, 3(1): 78–92.

Manandhar, N.P. and Manandhar, S. (2002) *Plants and People of Nepal*, Portland, OR: Timber Press.

Manor, J. (1998) "Southern Discomfort: The BJP in Karnataka," in T.B. Hansen and C. Jaffrelot (eds.) *The BJP and the Compulsions of Politics in India*, New Delhi: Oxford University Press.

Mansberger, J. (1987) "In Search of the Tree Spirit: Evolution of the Sacred Tree (Ficus Religiosa)," MA thesis, Department of Geography, University of Hawai'i, Honolulu.

March, K. (2002) *"If Each Comes Halfway": Meeting Tamang Women in Nepal*, Ithaca, NY: Cornell University Press.

Marcus, G.E. and Fischer, M.J. (1986) *Anthropology as Cultural Critique: An Experimental Moment in the Human Sciences*, Chicago: University of Chicago Press.

Marx, K. and Engels, F. (1970) *The German Ideology*, New York: International Publishers.

Mawdsley, E. (1998) "After Chipko: From Environment to Region in Uttarakhand," *Journal of Peasant Studies*, 25(4): 36–54.

—— (2004) "India's Middle Classes and the Environment," *Development and Change*, 35(1): 1073–97.

—— (2005) "The Abuse of Religion and Ecology: The Vishva Hindu Parishad and Tehri Dam," *Worldviews: Environment, Culture, Religion*, 8(2): 1–24.

—— (2006) "Hindu Nationalism, Postcolonialism and Environmental Discourses in India," *Geoforum*, 37(3): 380–90.

McKean, L. (1996) *Divine Enterprise: Gurus and the Hindu Nationalist Movement*, Chicago: University of Chicago Press.

Melanchthon, M. (2002) "Persecution of Indian Christians," *Dialog: A Journal of Theology*, 41(2): 103–13.

Messerli, B. and Hofer, T. (1995) "Assessing the Impact of Anthropogenic Land Use Change in the Himalayas," in G.P. Chapman and M. Thompson (eds.) *Water and the Quest for Sustainable Development in the Ganges Valley*, London: Mansell.

Messerschmidt, D.A. (1987) "Conservation and Society in Nepal: Traditional Forest Management and Innovative Development," in P.D. Little and M.M. Horowitz with A.E. Nyerges (eds.) *Lands at Risk in the Third World: Local-Level Perspectives*, Boulder, CO: Westview Press.

—— (1990) "Indigenous Environmental Management and Adaptation: An Introduction to Four Case Studies from Nepal," *Mountain Research and Development*, 10(1): 3–4.

Metz, J.J. (1990) "Forest Product Use in Upland Nepal," *Geographical Review*, 80(3): 279–87.

—— (1991) "A Reassessment of the Causes and Severity of Nepal's Environmental Crisis," *World Development*, 19(7): 805–20.

—— (1995) "Development in Nepal: Investment in the Status-Quo," *GeoJournal*, 35(2): 175–84.

Milton, K. (1996) *Environmentalism and Cultural Theory: Exploring the Role of Anthropology in Environmental Discourse*, London: Routledge.

—— (2002) *Loving Nature: Towards an Ecology of Emotion*, New York: Routledge.

Misra, A. (2000) "Hindu Nationalism and Minority Rights in India," *International Journal on Minority and Group Rights*, 7(1): 1–18.

Mitchell, T. (1999) "Rethinking the State: Genesis and Structure of the Bureaucratic

Field," in G. Steinmetz (ed.) *State/Culture: State Formation after the Cultural Turn*, Ithaca, NY: Cornell University Press.

Moddie, A.D. (1981) "Himalayan Environment," in J.S. Lall (ed.) *The Himalaya: Aspects of Change*, Bombay: Oxford University Press.

Molnar, A. (1981) *Nepal: The Dynamics of Traditional Systems of Forest Management: Implications for Community Forestry Development Policy*, Washington, DC: The World Bank.

Morris, J. ([1933] 1985) *The Gurkhas: An Ethnology*, Delhi: B. R. Publications.

Moseley, W. (2005) "Global Cotton and Local Environmental Management: The Political Ecology of Rich and Poor Small-Hold Farmers in Southern Mali," *Geographical Journal*, 171(1): 36–55.

Mosko, M.S. (1987) "The Symbols of 'Forest': A Structural Analysis of Mbuti Culture and Social Organization," *American Anthropologist*, 89(4): 896–913.

Müller-Böker, U. (1991) "Knowledge and Evaluation of the Environment in Traditional Societies of Nepal," *Mountain Research and Development*, 11(2): 101–14.

—— (1999) *The Chitawan Tharus of Southern Nepal: An Ethnoecological Approach*, Stuttgart: Franz Steiner Verlag.

Müller-Böker, U. and Kollmair, M. (2000) "Livelihood Strategies and Local Perceptions of a New Nature Conservation Project in Nepal," *Mountain Research and Development*, 20(4): 324–31.

Mumme, P.Y. (1998) "Models and Images for Vaisnava Environmental Theology: The Potential Contribution of Srivaisnavism," in L.E. Nelson (ed.) *Purifying the Earthly Body of God*, New York: SUNY Press.

Murashko, O. (2000) "The Concept of an International Ethnoecological Refuge," in P.P. Schweitzer, M. Biesele and R. K. Hitchcock (eds.) *Hunters and Gatherers in the Modern World: Conflict, Resistance, and Self-Determination*, New York: Berghahn Books.

Myers, N. (1986) "Environmental Repercussions of Deforestation in the Himalayas," *Journal of World Forest Resource Management*, 2(11): 63–72.

Nag, M., White, B.N.F. and Peet, R.C. (1978) "An Anthropological Approach to the Study of the Economic Value of Children: Java and Nepal," *Current Anthropology*, 19(2): 293–306.

Nagarajan, V. (2000) "Rituals of Embedded Ecologies: Drawing Kolams, Marrying Trees and Generating Auspiciousness," in C.K. Chapple and M.E. Tucker (eds.) *Hinduism and Ecology*, New Delhi: Oxford University Press.

Nanda, M. (2002) *Breaking the Spell of Dharma and Other Essays*, New Delhi: Three Essays Press.

—— (2003) *Prophets Facing Backwards: Postmodern Critiques of Science and Hindu Nationalism in India*, New Brunswick, NJ: Rutgers University Press.

Narayanan, V. (1997) " 'One Tree is Equal to Ten Sons': Hindu Responses to the Problems of Ecology, Population, and Consumption," *Journal of the American Academy of Religion*, 65(2): 291–332.

—— (2001) "Water, Wood and Wisdom: Ecological Perspectives from the Hindu Traditions," *Daedalus*, 130(4): 179–205.

Nelson, L.E. (ed.) (1998a) *Purifying the Earthly Body of God: Religion and Ecology in Hindu India*, New York: SUNY Press.

—— (1998b) "The Dualism of Nondualism: Advaita Vedanta and the Irrelevance of Nature," in L.E. Nelson (ed.) *Purifying the Earthly Body of God*, New York: SUNY Press.

—— (2000) "Reading the *Bhagavadgita* from an Ecological Perspective," in C.K. Chapple and M.E. Tucker (eds.) *Hinduism and Ecology*, New Delhi: Oxford University Press.

Nepal Development Forum (2002) *Medium Term Expenditure Framework 2002–5*, Kathmandu, Nepal: His Majesty's Government.

Nepal-'Yatri', P.P. (1983) *Rauté Lokajivana* [Rauté Folk Life], Kathmandu: HMG Ministry of Communication, Department of Information.

Nichter, M. (1981) "Negotiation of the Illness Experience: The Influence of Ayurvedic Therapy on the Psychosocial Dimension of Illness," *Culture, Medicine and Psychiatry*, 5(1): 5–24.

—— (2001) "The Political Ecology of Health in India: Indigestion as Sign and Symptom of Defective Modernization," in L.H. Connor and G. Samuel (eds.) *Healing Powers and Modernity: Traditional Medicine, Shamanism, and Science in Asian Societies*, Westport, CT: Bergin and Garvey.

Nield, R.S. (1985) "Fuelwood and Fodder: Problems and Policy," Working Paper, Water and Energy Commission Secretariat. Kathmandu, Nepal.

Nightingale, A.J. (2002) "Participating or Just Sitting In? The Dynamics of Gender and Caste in Community Forestry," *Journal of Forestry and Livelihood*, 2(1): 17–24.

—— (2003a) "A Feminist in the Forest: Situated Knowledges and Mixing Methods in Natural Resource Management," *ACME: An International E-Journal for Critical Geographers*, 2(1): 77–90.

—— (2003b) "Nature-Society and Development: Social, Cultural and Ecological Change in Nepal," *Geoforum*, 34(4): 525–40.

—— (2005) " 'The Experts Taught Us All We Know': Professionalisation and Knowledge in Nepalese Community Forestry," *Antipode*, 34(3): 581–604.

—— (2006) "The Nature of Gender: Work, Gender and Environment," *Environment and Planning D: Society and Space*, 24(2): 165–85.

Norval, A.J. (2001) "Reconstructing National Identity and Renegotiating Memory: The Work of the TRC," in T.B. Hansen and F. Stepputat, *States of Imagination: Ethnographic Explorations of the Postcolonial State*, Durham, NC: Duke University Press.

Odum, E.P. (1969) "The Strategy of Ecosystem Development," *Science*, 164: 262–70.

Oliver, C.D. (1981) "Forest Development in North America Following Major Disturbances," *Forest Ecology and Management*, 3(3): 153–68.

Olson, M.D. (1997) "Re-constructing Landscapes: The Social Forest, Nature and Spirit-World in Samoa," *Journal of the Polynesian Society*, 106(1): 7–32.

Osella, F. and Osella, C. (2000) *Social Mobility in Kerala: Modernity and Identity in Conflict*, London: Pluto Press.

Ostrom, E. (1990) *Governing the Commons: The Evolution of Institutions for Collective Action*, Cambridge: Cambridge University Press.

Overseas Development Agency (1997) *Natural Resources Research: Working for Development, Third Report on ODA's Renewable Research Strategy*, London: ODA Natural Resources Development Department.

Owens, B. (2002) "Monumentality, Identity, and the State: Local Practice, World Heritage, and Heterotopia at Swayambhu, Nepal," *Anthropological Quarterly*, 75(2): 269–316.

Palmer, M. (1996) *Travels Through Sacred China*, London: HarperCollins.

Palmer, M. and Finlay, V. (2003) *Faith in Conservation: New Approaches to Religions and the Environment*, Washington, DC: World Bank.

Pálsson, G. (1996) "Human–Environmental Relations: Orientalism, Paternalism and Communalism," in P. Descola and G. Pálsson (eds.) *Nature and Society: Anthropological Perspectives*, London: Routledge.

Pande, P.C., Joshi, G.C. and Kandpal, M.M. (1990) "Ethnobotany of Kumaun Himalaya," in N.K. Sah, S.D. Bhatt and R.K. Pande (eds.) *Himalaya: Environment, Resources and Development*, Almora: Shree Almora Book Depot.

Pandey, B.P. (1989) *Sacred Plants of India*, New Delhi: Shree Publishing House.

Pandey, R.N. (1997) *Making of Modern Nepal: A Study of History, Art, and Culture of the Principalities of Western Nepal*, New Delhi: Nirala Publications.

Parajuli, P. (2001) "No Nature Apart: Adivasi Cosmovision and Ecological Discourses in Jharkhand," in P. Arnold and A.G. Gold (eds.) *Sacred Landscape and Cultural Politics: Planting a Tree*, Aldershot: Ashgate Publishing.

Parker, B. (1988) "Ritual Coordination of Medical Pluralism in Highland Nepal: Implications for Policy," *Social Science and Medicine*, 27(9): 919–25.

Patton, L.L. (2000) "Nature Romanticism and Sacrifice in the Rgvedic Interpretation," in C.K. Chapple and M.E. Tucker (eds.) *Hinduism and Ecology*, New Delhi: Oxford University Press.

Pedersen, P. (1995) "Nature, Religion and Cultural Identity: The Religious Environmentalist Paradigm," in O. Bruun and A. Kalland (eds.) *Asian Perceptions of Nature: A Critical Approach*, Richmond: Curzon Press.

Peet, R. and Watts, M. (2004) *Liberation Ecologies: Environment, Development, Social Movements*, London, Routledge.

Peluso, N. and Watts, M. (2001) *Violent Environments*, Ithaca, NY: Cornell University Press.

Pemberton, J. (1994) *On the Subject of Java*, Ithaca, NY: Cornell University Press.

Peters, P.E. (1996) "Introduction: 'Who's Local Here?': The Politics of Participation in Development," *Cultural Survival Quarterly*, 20(3): 22–5.

Pickett, S.T. and White, P.S. (1985) *The Ecology of Natural Disturbance and Patch Dynamics*, Orlando, FL: Academic Press.

Pigg, S.L. (1992) "Inventing Social Categories through Place: Social Representations and Development in Nepal," *Comparative Studies in Society and History*, 34(3): 491–513.

—— (1993) "Unintended Consequences: The Ideological Impact of Development in Nepal," *South Asia Bulletin*, 13(1&2): 45–58.

—— (1996) "The Credible and the Credulous: The Question of 'Villagers' Beliefs' in Nepal," *Cultural Anthropology*, 11(2): 160–201.

Pitt, D.C. (1986) "Crisis, Pseudocrisis, or Supercrisis? Poverty, Women, and Young People in the Himalaya: A Survey of Recent Developments," *Mountain Research and Development*, 6(2): 119–31.

Poffenberger, M. (1980) *Patterns of Change in the Nepal Himalaya*, Delhi: Macmillan India.

Pokhrel, B., Tripathi, V., Dahal, B., Parajuli, K., Sharma, G. and Bhattarai, H. (2004) V.S. (1983–84 A.D.). *Nepālī Bṛihat Śabdakoś*. Kathmandu: Nepāli Rājakiya Prajna-Pratiṣṭhān.

Post, M. (2003) "Kathmandu's Malignant Urban Tumor," *Nepali Times* #161, September 6–11.

Post Reporter (2000) "Billions Spent but the River Still Polluted," *Kathmandu Post*, February 7.

Pottier, J. (2003) "Introduction," in J. Pottier, A. Bicker and P. Sillitoe (eds.) *Negotiating Local Knowledge: Power and Identity in Development*, London: Pluto Press.

Prime, R. (1992) *Hinduism and Ecology: Seeds of Truth*, London and New York: Cassell.

Proksch, A. (1995) "Foreword," in A. Proksch and H.R. Baidya (eds.) *Images of a Century: The Changing Townscapes of the Kathmandu Valley*, Kathmandu, Nepal: Deutsche Gesellschaft für Technische Zusammenarbeit: Urban Development Through Local Efforts Project.

Rademacher, A.M. (2005) "Culturing Urban Ecology: Development, Statemaking, and River Restoration in Kathmandu," PhD thesis, Yale University.

Rademacher, A.M. and Patel, R. (2002) "Retelling Worlds of Poverty: Reflections on Transforming Participatory Research for a Global Narrative," in K. Brock and R. McGee (eds.) *Knowing Poverty: Critical Reflections on Participatory Research and Policy*, London: Earthscan.

Raeper, W. and Hoftun, M. (1992) *Spring Awakening: An Account of the 1990 Revolution in Nepal*, Delhi: Viking.

Rajalakshmi, T.K. (2001) "A Saffron Twist," *Frontline*, 18(8), 14–27 April. Online. Available at: www.hinduonnet.com/fline/fl1808/18080420.htm.

Ramble, C. (1995) "Gaining Ground: Representations of Territory in Bon and Tibetan Popular Tradition," *Tibet Journal*, 20: 83–124.

Rambo, A.T. (1985) *Primitive Polluters: Semang Impact on the Malaysian Tropical Rain Forest Ecosystem*, Anthropological Papers No. 76, Ann Arbor, MI: Museum of Anthropology, University of Michigan.

Ramirez, P. (2000) *De la Disparition des Chefs: une anthropologie politique népalaise*. Collection Monde Indien, Paris: CNRS éditions.

Ramsay, W.J.H. (1986) "Erosion Problems in the Nepal Himalaya: An Overview," in S.C. Joshi *et al.* (eds.) *Nepal Himalaya Geo-ecological Perspectives*, Naini Tal, India: Himalayan Research Group.

Rangan, H. (2000a) "State Economic Policies and Changing Regional Landscapes in the Uttarakhand Himalaya, 1818–1947," in A. Agrawal and K. Sivaramakrishnan (eds.) *Agrarian Environments*, Durham, NC: Duke University Press.

—— (2000b) *Of Myths and Movements: Rewriting Chipko into Himalayan History*, New Delhi: Oxford University Press.

Rappaport, R.A. (1979) "On Cognized Models," in R.A. Rappaport (ed.) *Ecology, Meaning and Religion*, Berkeley, CA: North Atlantic Books.

Redman, C.L. (1999) *Human Impact on Ancient Environments*, Tucson, AZ: University of Arizona Press.

Regmi, M.C. (1971) *A Study in Nepali Economic History, 1768–1846* Bibliotheca Himalayica, ser. 1, v. 14. New Delhi: Manjuśrī Publishing House.

—— (1978) *Thatched Huts and Stucco Palaces: Peasants and Landlords in 19th-Century Nepal*, New Delhi: Vikas.

—— (1988) *An Economic History of Nepal 1846–1901*, Varanasi: Nath Publishing House.

Reichel-Dolmatoff, G. (1976) "Cosmology as Ecological Analysis: A View from the Rain Forest," *Man* (New Series), 11(3): 307–18.

Risley, H.H. ([1891] 1981) *The Tribes and Castes of Bengal: Ethnographic Glossary*, Vol. 2, Calcutta: Firma Mukhopadhyay.

Robbe, E. (1954) *Report to the Government of Nepal on Forestry*, ETAP Report No. 209, Rome: Food and Agricultural Organization of the United Nations.

Rodman, M. (1992) "Empowering Place: Multilocality and Multivocality," *American Anthropologist*, 94(3): 640–56.

Roe, E. M. (1991) "Development Narratives, or Making the Best of Blueprint Development," *World Development*, 19(4): 287–300.

—— (1995) "Except Africa: Postscript to a Special Section on Development Narratives," *World Development*, 23(6): 1065–9.

Roy, R. (1993) "Swadhyaya: Values and Message," in P. Wignaraja (ed.) *New Social Movements in the South: Empowering People*, London: Zed Books.

Roy, S.C. ([1925] 1978) *The Birhors: A Little-Known Jungle Tribe of Chota Nagpur*, Ranchi: Man in India Office.

Rukmani, T.S. (2000) "Literary Foundations for an Ecological Aesthetic: *Dharma*, Ayurveda, the Arts, and Abhijnanaskuntalam," in C.K. Chapple and M.E. Tucker (eds.) *Hinduism and Ecology*, New Delhi: Oxford University Press.

Russell, A.J. (1992) "The Hills are Alive with the Sense of Movement: Migration and Identity amongst the Yakha of East Nepal," *Himalayan Research Bulletin*, 12(1–2): 35–43.

—— (2000) "Breaking Bones: Political Eruptions and Social Ruptures in Nepal, 1989–90," in S. Basu and S. Das (eds.) *Electoral Politics in South Asia*, University of Calcutta, Department of History Monograph 18, Calcutta: K.P. Bagchi and Co.

Saberwal, V. (2000) "Environmental Alarm and Institutionalized Conservation in Himachal Pradesh, 1865–1994," in A. Agrawal and K. Sivaramakrishnan (eds.) *Agrarian Environments*, Durham, NC: Duke University Press.

Sagant, P. (1969) "Tâmpuńmâ, Divinité Limbu de la Forêt," *Objets et Mondes* 9: 107–24; reprinted in *The Dozing Shaman: The Limbus of Eastern Nepal*, Delhi: Oxford University Press.

—— (1976) *Le Paysan Limbu: Sa Maison et Ses Champs*, Paris: Mouton.

—— (1996) *The Dozing Shaman: The Limbus of Eastern Nepal*, Delhi: Oxford University Press.

Samuel, R. (1994) *Theatres of Memory: Past and Present in Contemporary Culture*, London: Verso.

Saraswati, B. (1998) "The *Logos* and the *Mythos* of the Sacred Grove," in P.S. Ramakrishnan, K.G. Saxena and U.M. Chandrashekar (eds.) *Conserving the Sacred for Biodiversity Management*, Enfield, NH: Science Publishers, Inc.

Sarkar, S. (2002) *Beyond Nationalist Frames: Postmodernism, Hindu Fundamentalism, History*, New Delhi: Permanent Black.

Sax, W.S. (2003) "Divine Kingdoms in the Central Himalayas," in N. Gutschow *et al.* (eds.) *Sacred Landscape of the Himalaya: Proceedings of an International Conference at Heidelberg*, Vienna: Austrian Academy of Sciences Press.

Sayer, A. (1993) "Postmodernist Thought in Geography: A Realist View," *Antipode*, 25(4): 320–44.

Schrader, H. (1988) *Trading Patterns in the Nepal Himalayas*, Saarbrucken: Verlag Breitenbach.

Schroeder, R.F. (1985) "Himalayan Subsistence Systems: Indigenous Agriculture in Rural Nepal," *Mountain Research and Development*, 5(1): 31–44.

Scott, J.C. (1985) *Weapons of the Weak: Everyday Forms of Peasant Resistance*, New Haven, CT: Yale University Press.

—— (1998) *Seeing Like a State: How Certain Schemes to Improve the Human Condition Have Failed*, New Haven, CT: Yale University Press.

Seddon, D. (1987) *Nepal: A State of Poverty*, New Delhi: Vikas Publishing House.
Senior, H.R. (1977) *A Vocabulary of the Limbu Language of Eastern Nepal*, Kathmandu: Ratna Pustak Bhandar.
Sharma, A. (1998) "Attitudes to Nature in the Early Upanishads," in L.E. Nelson (ed.) *Purifying the Earthly Body of God*, New York: SUNY Press.
Sharma, B. (2000) "My God! What a Big Drain!" *Kathmandu Post*, 24 August.
Sharma, M. (2001a) "Nature and Nationalism," *Frontline*, 18(3). February 3–16. Online. Available at: www.hinduonnet.com/fline/fl1803/18030940.htm.
—— (2001b) *Landscapes and Lives: Environmental Dispatches on Rural India*, New Delhi: Oxford University Press.
—— (2002) "Saffronising Green," *Seminar* (New Delhi), 516: 26–30. Online. Available at: www.india-seminar.com/2002/516/516%20mukul%20sharma.htm.
Sharma, S. ([1929] 1983) *The System of Ayurveda*, New Delhi: Neeraj Publishing House.
—— (2002) "The Hindu State and the State of Hinduism," in K.M. Dixit and S. Ramachandaran (eds.) *State of Nepal*, Lalitpur: Himal Books.
Shengi, P. (ed.) (1996) *Banking on Biodiversity: Report on the Regional Consultation on Biodiversity Assessment in the Hindu-Kush Himalayas*, Kathmandu: ICIMOD.
Sherma, R.D. (1998) "Sacred Immanence: Reflections of Ecofeminism in Hindu Tantra," in L.E. Nelson (ed.) *Purifying the Earthly Body of God*, New York: SUNY Press.
Shields, R. (1991) *Places on the Margin: Alternative Geographies of Modernity*, London: Routledge.
Shiva, V. (1989) *Staying Alive: Women, Ecology and Development*, London: Zed Books.
Shore, B. (1989) "Mana and Tapu," in A. Howard and R. Borofsky (eds.) *Developments in Polynesian Ethnology*, Honolulu: University of Hawai'i Press.
Shrestha, N (1995) "Becoming a Development Category," in J.S. Crush (ed.) *Power of Development*, London: Routledge.
Singh, M. (1992) *Displacement by the Sardar Sarovar and Tehri: A Comparative Study of Two Dams*, Bombay: Multiple Action Research Group (MARG).
Singh, N.B. (1997) *The Endangered Rauté Tribe: Ethnobiology and Biodiversity*, Kathmandu: Global Research Carrel for Ethnobiology.
Singh, S.P., Adikari, B.S. and Zobel, D.B. (1994) "Biomass, Productivity, Leaf Longevity, and Forest Structure in the Central Himalaya," *Ecological Monographs*, 64(4): 401–21.
Sinha, A.C. (1975) *Politics of Sikkim: A Sociological Study*, Faridabad: Thompson Press.
Sinha, R.K. (1995) "Biodiversity Conservation Through Faith and Tradition in India – Some Case Studies [Sacred Groves]," *International Journal of Sustainable Development and World Ecology*, 2(4): 278–84.
Sinha, S. (1962) "State Formation and Rajput Myth in Tribal Central India," *Man in India*, 42(1): 35–80.
Sinha, S., Greenberg, B. and Gururani, S. (1997) "The 'New Traditionalist' Discourse of Indian Environmentalism," *Journal of Peasant Studies*, 24(3): 65–99.
Siu, H. (1990) "Recycling Tradition: Culture, History, and Political Economy in the Chrysanthemum Festivals of South China," *Comparative Studies in Society and History*, 32(4): 765–94.
Sivaramakrishnan, K. (1999) *Modern Forests: Statemaking and Environmental Change in Colonial Eastern India*, Oxford: Oxford University Press.

Skaria, A. (1999) *Hybrid Histories: Forests, Frontiers and Wildness in Western India*, Delhi: Oxford.

Slusser, M. (1982) *Nepal Mandala: A Cultural Study of the Kathmandu Valley*, Princeton, NJ: Princeton University Press.

Sperber, D. (2006) "Anthropology and Psychology: Towards an Epidemiology of Representations," in H.L. Moore and T. Sanders (eds.) *Anthropology in Theory: Issues in Epistemology*, Malden, MA: Blackwell Publishing.

Sponsel, L.E., Natadecha, P., Ruttanadakul, N. and Juntadach, S. (1998) "Sacred and/or Secular Approaches to Biodiversity Conservation in Thailand," *Worldviews: Environment, Culture, Religion*, 2(2): 155–67.

Stanley International (1994) *Bagmati Basin Water Management Strategy and Investment Plan: Final Report*, Kathmandu: HMG/MHHP and the World Bank/Japanese Grant Fund.

Starkel, L. (1972) "The Role of Catastrophic Rainfall in the Shaping of the Relief of the Lower Himalaya (Darjeeling Hills)," *Geographica Polonica*, 21: 103–47.

Steinmann, B. (2001) *Les Enfants du Singe et de la Démone: Mémoires des Tamang récits Himalayens*, Nanterre: Société d'ethnologie.

Sterling, C. (1976) "Nepal," *The Atlantic Monthly*, 238(4): 14–25.

—— (1981) *The Terror Network: The Secret War of International Terrorism*, New York: Henry Holt and Co.

Stevens, S.F. (1993) *Claiming the High Ground: Sherpas, Subsistence, and Environmental Change in the Highest Himalaya*, Berkeley: University of California Press.

Stiller, L.F. (1975) *The Rise of the House of Gorkha (1768–1816)*, Kathmandu: Ratna Pustak Bhandar.

—— (1976) *The Silent Cry: The People of Nepal 1816–1839*, Kathmandu: Sahayogi Press.

Stocking, M. (1996) "Soil Erosion: Breaking New Ground," in M. Leach and R. Mearns (eds.) *The Lie of the Land*, Oxford: International African Institute.

Subba, J.R. (2002) *History and Development of Limboo Language*, Gangtok: Sukhim Yakthung Mundhum Saplopa.

Subba, S. (2000) "Bagmati Waits Even as Government Ignores Cleanup Call," *Kathmandu Post*, 10 June.

Subba, T.B. (1990) "Limbu Ethnicity at Crossroads," *Himalayan Miscellany*, 4: 40–45.

—— (1999a) "Limbu Nationalism and Integration," in A.K. Danda (ed.) *Ethnicity, Nationalism and Integration*, Calcutta: The Asiatic Society.

—— (1999b) *Politics of Culture: A Study of Three Kirata Communities in the Eastern Himalayas*, Hyderabad: Orient Longman.

Sullivan, B.M. (1998) "Theology and Ecology at the Birthplace of Kṛṣṇa," in L.E. Nelson (ed.) *Purifying the Earthly Body of God*, New York: SUNY Press.

Sutherland, P. (2005) "T(r)opologies of Rule (Raj): Ritual Sovereignty and Theistic Subjection," *European Bulletin of Himalayan Research*, 29–30: 82–119.

Swyngedouw, E. (1997) "Neither Global nor Local: 'Glocalization' and the Politics of Scale," in K. Cox (ed.) *Spaces of Globalization: Reasserting the Power of the Local*, New York: The Guilford Press.

Taylor, B. and Kaplan, J. (2003) *Encyclopedia of Religion and Nature*, New York: Continuum International.

Thapa, D. and Sijapati, B. (2005) *A Kingdom Under Siege: Nepal's Maoist Insurgency 1996–2004*, New York: Zed Books.

Thapa, P. (1989) *Nepal: Socio-economic Change and Rural Migration*, Delhi: Vikas Publishing House.

Theodossopoulos, D. (2003) *Troubles with Turtles: Cultural Understandings of the Environment on a Greek Island*, New York: Berghahn Books.

Thomas, D. and Middleton, N. (1994) *Desertification: Exploding the Myth*, Chichester: John Wiley and Sons.

Thomas, K. (1983) *Man and the Natural World: Changing Attitudes in England, 1500–1800*, London: Allen Lane.

Thompson, M. and Warburton, T. (1985) "Uncertainty on a Himalayan Scale," *Mountain Research and Development*, 5(2): 115–35.

Thompson, M., Warburton, M. and Hatley, T. (1986) *Uncertainty on a Himalayan Scale: An Institutional Theory of Environmental Perception and a Strategic Framework for the Sustainable Development of the Himalaya*, London: Milton Ash Editions/Ethnographica.

Times of India (2000) "VHP sees Tehri Dam as a Threat to Hinduism," 14 July.

Tiwari, P. C. (2000) "Land-Use Changes in Himalaya and Their Impact on the Plains Ecosystem: Need for Sustainable Land Use," *Land Use Policy*, 17(2): 101–11.

Toffin, G. (1990) "L'Espace des Populations Montagnards du Népal: Le Cas des Tamang du Ganesh Himal," in J. Bourliaud, J-F. Dobremez and F. Vigny (eds.) *Sociétés Rurales des Andes et de l'Himalaya*, Grenoble: Versants.

Tomalin, E. (2002) "The Limitations of Religious Environmentalism for India," *Worldviews: Environment, Culture, Religion*, 6(1): 12–30.

Tuan, Y.F. (1977) *Space and Place*, Minneapolis: University of Minnesota Press.

—— (1978) "Sacred Space: Exploration of an Idea," in K.W. Butzer (ed.) *Dimensions of Human Geography: Essays on Some Familiar and Neglected Themes*, Research Paper 186, Chicago: University of Chicago, Department of Geography.

—— (1979) *Landscapes of Fear*, New York: Pantheon Books.

Tucker, M.E. (1997) "The Emerging Alliance of Religion and Ecology," *Worldviews: Environment, Culture, Religion*, 1(1): 3–24.

Tuladhar, J.M. (1977) *The Population of Nepal: Structure and Change*, Berkeley: University of California Press.

Tulku, R. (1991) "Pre-Buddhistic Rites in Buddhism," in T.B. Subba and K. Datta (eds.) *Religion and Society in the Himalayas*, New Delhi: Gian Publishing House.

Turnbull, C. (1961) *The Forest People*, New York: Simon & Schuster.

Turner, R.L. ([1931] 1980) *A Comparative and Etymological Dictionary of the Nepali Language*, New Delhi: Allied Publishers Pvt Ltd.

Upreti, B.P. (1976) "Limbuan Today: Process and Problems," *Contributions to Nepalese Studies*, 3(2): 47–70.

Vansittart, E. (1915) *Gurkhās*, Calcutta: Superintendent of Government Printing, India.

Varma, P.K. (1998) *The Great Indian Middle Class*, New Delhi: Penguin.

Virgo, K.J. and Subba, K.J. (1994) "Land-Use Change Between 1978 and 1990 in Dhankuta District, Koshi Hills, Eastern Nepal," *Mountain Research and Development*, 14(2): 159–70.

Visvader, J. (1986) "Philosophy and Human Ecology," in R.J. Borden (ed.) *Human Ecology: A Gathering of Perspectives*, College Park, OH: The Society for Human Ecology.

Viveiros de Castro, E. (1996) "Images of Nature and Society in Amazonian Ethnology," *Annual Reviews of Anthropology*, 25: 179–200.

—— (1998) "Cosmological Deixis and Amerindian Perspectivism," *Journal of the Royal Anthropological Institute*, 4(3): 469–88.

Wallace, M. (1981) "Solving Common Property Resource Problems: Deforestation in Nepal," PhD thesis, Harvard University.

Weinberg, A. (1972) "Science and Trans-Science," *Minerva*, 10(2): 209–22.

Whatmore, S. (2002) *Hybrid Geographies: Natures, Cultures, Spaces*, Thousand Oaks, CA: Sage.

Whelpton, J. (2005) *A History of Nepal*, New York: Cambridge University Press.

White, D. (1996) *The Alchemical Body*, Chicago: University of Chicago Press.

White, L., Jr. (1967) "The Historical Roots of Our Ecological Crisis," *Science*, 155: 48–56.

Wiart, J. (1983) "Ecosystème Villageois Traditionnel en Himalaya Népalais: La Production Forestière Suffit-Elle aux Besoins de la Population?" PhD thesis, Université Joseph Fourier, Grenoble, France.

Williams, R. (1961) *Culture and Society 1780–1950*, London: Penguin Books.

—— (1980) "Ideas of Nature," in *Problems in Materialism and Culture*, London: Verso.

WRI (1995) *Tropical Forests: A Call for Action. Report of an International Task Force Convened by the World Resources Institute, the World Bank, and the United Nations Development Programme*, Washington, DC: World Resources Institute.

Wu, K. and Thornes, J.B. (1995) "Terrace Irrigation of Mountainous Hill Slopes of the Middle Hills of Nepal: Stability or Instability?," in G.P. Chapman and M. Thompson (eds.) *Water and the Quest for Sustainable Development in the Ganges Valley*, London: Mansell.

Wujastyk, D. (2003) *The Roots of Ayurveda*, London: Penguin Books.

WWF (1986) *The Assisi Declaration; Messages on Man and Nature from Buddhism, Christianity, Hinduism, Islam and Judaism*, London: World-Wide Fund for Nature.

Wyatt-Smith, J. (1982) *The Agricultural System in the Hills of Nepal: Ratio of Agricultural to Forest Land and the Problem of Animal Fodder*, Occasional Paper No.1, Kathmandu: Agricultural Projects Services Centre (APROSC).

Zavos, J. (2000) *The Emergence of Hindu Nationalism in India*, Oxford: Oxford University Press.

Zimmerer, K. (1994) "Human Geography and the 'New Ecologies': The Prospect and Promise of Integration," *Annals of the Association of American Geographers*, 84(1): 108–25.

Zimmermann, F. (1987) *The Jungle and the Aroma of Meats: An Ecological Theme in Hindu Medicine*, Berkeley: University of California Press.

Zomer, R.J., Ustin, S.L. and Carpenter, C.C. (2001) "Land Cover Change along Tropical and Subtropical Riparian Corridors within the Makalu Barun National Park and Conservation Area, Nepal," *Mountain Research and Development*, 21(2): 175–83.

Zurick, D.N. (1990) "Traditional Knowledge and Conservation as a Basis for Development in a West Nepal Village," *Mountain Research and Development*, 10(1): 23–33.

Zurick, D.N. and Karan, P.P. (1999) *Himalaya: Life on the Edge of the World*, Baltimore: Johns Hopkins University Press.

Zysk, K. (1991) *Asceticism and Healing in Ancient India: Medicine in the Buddhist Monastery*, New York: Cambridge University Press.

Index

For Product Safety Concerns and Information please contact our EU representative GPSR@taylorandfrancis.com
Taylor & Francis Verlag GmbH, Kaufingerstraße 24, 80331 München, Germany

www.ingramcontent.com/pod-product-compliance
Ingram Content Group UK Ltd.
Pitfield, Milton Keynes, MK11 3LW, UK
UKHW011000180926
475701UK00004B/486
9780415533140